"Look Here, Sir, What a Curious Bird"

SEARCHING FOR ALI, ALFRED RUSSEL WALLACE'S FAITHFUL COMPANION

PAUL SPENCER SOCHACZEWSKI

EXPLORER'S EYE PRESS

GENEVA, SWITZERLAND

For Lesley,
For Dic
w
of our 1
admiration for
your energy
in discovering
new ways
of looking
at the magic
of Asia.
Continue to
travel well, with
clear eyes and
good heart.
Paul
London
January 2024

About the Cover

Credits:
Ali: Trustees of the Natural History Museum, London
Wallace standardwing: Hand-colored lithograph of *Semioptera wallacii* by John and Elizabeth Gould from John Gould's *The Birds of Australia*. Trustees of the Natural History Museum, London

Ali (family name unknown), age about 22, and the bird-of-paradise he shot and presented to Alfred Russel Wallace.

The creature — now known as *Semioptera wallacii*, or Wallace's standardwing — turned out to be one of the 212 new bird species Wallace collected; this was his only new bird-of-paradise.

The book's title refers to this incident, about which Wallace quotes Ali as saying (likely translated from Malay): "Look here, sir, what a curious bird." It is one of just two times Wallace quotes Ali directly.

We know quite a bit about Alfred Russel Wallace, one of the great figures of modern science.

But what of Ali, Wallace's faithful companion who supported him during much of his eight-year sojourn in the Malay Archipelago? What did teenage Ali think of this tall, strange European man who swooned over a new beetle, cuddled a baby orangutan he had orphaned, and lived rough in the forest? What did Wallace teach Ali? What did Ali teach Wallace? Where did Ali go after Wallace left? Can we find Ali's descendants to spur a conservation movement? Could I speak with Ali's spirit through a medium? These questions have sparked my interest for some 50 years.

Praise for
"Look Here, Sir, What a Curious Bird"

"Once again, Paul Spencer Sochaczewski has created a new literary genre, which he simply calls an 'enhanced biography' of Wallace's assistant Ali. His creativity is remarkable – he somehow manages to combine solid scholarship, personal memoir, good heart and hearty humor, delicious speculation, soul-touching imagined scenarios, a plenitude of cartoons and photos, and riveting asides. He even invokes the aid of mediums and shamans to pursue his quest to learn more about this intriguing and under-appreciated 19th-century Borneo teenager. A splendid work of innovative history writing."

– James Clad, Board of Supervisors, Library of Congress, and former professor of Asian Studies, Georgetown University

"Through the ages, our famous explorers and adventurers have rarely pursued their quests alone. They've been helped along by others who, barely acknowledged, have helped them on their way through seemingly impenetrable forests, swamps, and ravines. This was the case with Alfred Russel Wallace, the great Victorian naturalist whose daring expeditions were aided by a young man referred to simply as 'Ali.'

This is an important, enlightening book not only because it tells Ali's story, at last giving him the prominence he deserves, but because it reminds us that we each have Alis in our lives, people to whom we should from time to time raise a grateful glass in thanks."

– Benedict Allen, BBC Television explorer and author of Explorer: The Quest for Adventure and the Great Unknown

"The majority of the bird specimens collected by Alfred Russel Wallace during his time in the Malay Archipelago are kept at the Natural History Museum at Tring, UK. But while Wallace gets the credit for compiling his impressive collection, we should realize that some 5,000 of Wallace's total of 8,050 birds were actually collected, and often skinned and prepared, too, by his assistant Ali. Without their skilled and dedicated assistants, the famous men and women of science would not have been as successful as they were. Paul Spencer Sochaczewski's insightful and gloriously written book shines the much-deserved light on Ali, who was such an important but sadly overlooked helper to Wallace."

— *Hein van Grouw, senior curator, Bird Group, The Natural History Museum, UK*

"'Who am I?' is a question we sometimes ask ourselves without ever coming up with a definitive answer. Our stories change according to age, circumstance, and experience. How much harder then to ask the same question of an unheralded 19th-century teenager from Borneo. Paul Spencer Sochaczewski approaches Ali, Alfred Russel Wallace's loyal assistant, from every conceivable angle, bringing to bear academic research, colonial perceptions, and the mystical mindset that would have been Ali's own to define the kaleidoscopic facets of his being. Fieldworkers know the crucial importance of the — usually nameless — 'local assistant' who significantly helped the scholar's research but whose name never made it into print. This important and delightful work questions both the conventional limits of biography and highlights an often-ignored element that is crucial to the development of the myth of 'Western scientific achievement.'"

— *Nigel Barley, author of* White Rajah *and* The Innocent Anthropologist, *and former curator for Southeast Asia, British Museum*

"Readers of his earlier books have been enthralled by Paul Spencer Sochaczewski's perceptions of people and places of the island realm that was named 'the Malay Archipelago' by the 19th-century natural history collector and naturalist, Alfred Russel Wallace. Paul has now focused his attention on the Malay youth, recruited by Wallace in Sarawak — a man, named only as Ali, whose features were captured by a single photograph. Not one of Wallace's other temporary assistants was as competent and none so loyal as Ali, who remained with him until the end of his journey in the Archipelago. As a collector, Wallace might not have succeeded as he did without Ali, whose competence and, at times, compassion, undoubtedly safe-guarded and supported Wallace during his travels among the many different islands now united under the red and white flag of Indonesia.

"An additional attraction of Paul's interpretive writing is his fascination with the mystical, in ways that still linger in the consciousness and practices of many thoughtful people in modern Indonesia. In this 'enhanced biography,' he has described in detail his search for the true identity of Ali through the intervention of dukuns, spiritual mediums. He offers few certainties but certainly provides readers with an imaginative perspective on the enigmatic personality of the man who probably retired to Ternate island and there styled himself 'Ali Wallace.'"

— *Dato Sri Gathorne, Lord Cranbrook, author of* Mammals of Borneo, Mammals of Southeast Asia, *and* Wonders of the Natural World of Southeast Asia

"Alfred Russel Wallace and Charles Darwin are often linked for their work in developing the Theory of Evolution by Natural Selection. But the career paths of the two men could hardly have been more different. Darwin sailed in relative comfort on an official British navy ship, while Wallace travelled independently through the Southeast Asian archipelago on perilous native craft. Darwin's experience on the Beagle has been well-chronicled; Wallace's much more adventurous voyages have attracted little attention.

"Darwin does not seem to have developed any important relationships with the people of the Pacific, while Wallace more or less adopted a Malay teenager named Ali, who shared his travels for years. Paul Spencer Sochaczewski skillfully gathers information about him as reflected in the words of others, mainly Wallace, to reconstruct something of the history of a boy/man who otherwise would have joined the extensive ranks of the 'people without history' but who were nonetheless directly responsible for shaping the way we think of ourselves and our relationship with nature today."

— *John Miksic, emeritus professor, National University of Singapore; senior research fellow, Nanyang Technological University; and author of* Borobudur: Majestic Mysterious Magnificent

"Like Alexander von Humboldt before him, Alfred Russel Wallace had struck out with most of the guides and assistants he hired during his South American travels and was not having better luck in his first year in the Malay Archipelago in the mid-1850s. Then he met the Malay teenager known only as Ali, and things turned around. Ali proved to be a most faithful companion and highly competent as an interpreter, guide, collector, and cook. The story of their relationship, as told here by journalist Paul Spencer Sochaczewski, is a most interesting one, enlivened as it is by the author's dedication to the subject, the relative dearth of information on Ali, and the reality that historical accounts are seldom clear and absolute but often a mash-up of speculation and morsels of information crafted into a feasible tale."

— *Charles H. Smith, founder and curator of The Alfred Russel Wallace Page*

"Another fascinating and insightful book from the prolific pen of Paul Spencer Sochaczewski. Some of us are lucky enough to have had an Ali in our lives. I have had two outstanding such individuals: Kelakua, the best *mehariste* in the Sahara, a Tuareg with whom I rode by camel many miles through the Niger desert; and Nyapun, my dearest friend, a nomadic Penan, with whom I travelled through the rainforest for a year in Borneo."

— *Robin Hanbury-Tenison, explorer and author of* Mulu: The Rain Forest, The Oxford Book of Exploration, *and* The Great Explorers

"Alfred Russel Wallace's assistant Ali is an adopted son of Ternate. He lived here for some three years in the mid-19th century, and 'retired' in Ternate after Wallace left for England. And it was in Ternate where Alfred Russel Wallace wrote his famous Ternate Essay that outlined his Theory of Evolution by Natural Selection. Paul Spencer Sochaczewski's story of Ali tells of an important period in Ternate's long and proud history. And his book will be an important stimulus to continue conservation efforts and to educate Ternate citizens about the important role the island has played in the history of science."

—Tauhid Soleman, mayor of Ternate, Indonesia

"North Maluku is one of the richest parts of the world for biological diversity. It was here that Alfred Russel Wallace made some of his most exciting and important discoveries — the famous Wallace's standardwing bird-of-paradise, that gives the book its title, was found on the North Maluku island of Bacan. This fascinating and informative book should be essential reading for everyone — both residents of North Maluku and visitors — who want to learn more about our province's nature, and by extension, the need to conserve, protect, and love what we have."

— Abdul Ghani Kasuba, governor of North Maluku Province, Indonesia

"I have known Paul Spencer Sochaczewski for years and worked with him and other like-minded friends to search for Ali's descendants in Ternate and the location of Wallace's house. As director of the Fort Oranje Historical Site, I appreciate how important it is to present history in an informative, and entertaining manner, as Sochaczewski has done. His new book, *"Look Here, Sir, What a Curious Bird,"* is a vital resource and strongly recommended for anyone who wants to visit Ternate and the region."

— Rinto Taib, director of the Spice Museum and Fort Oranje Historic Site, Ternate, Indonesia

"The achievements of Alfred Russel Wallace in Indonesia, assisted by the young man we know only as Ali, are important to the nation's past, but more important, are a source of inspiration and motivation for the younger generation. Wallace was British, but he was not a colonizer; he made no attempt to impose his views and beliefs on the 19th-century population of our country. Just the opposite — he traversed the country, from Sumatra to Papua, respected the people he met, and marveled at the country's astounding biological diversity and rich natural heritage. Studying, conserving, and being proud of our flora and fauna is important to modern-day Indonesia. Our heritage is found not only in ancient temples and statues but also as living, breathing wildlife in our forests and oceans — our Tanah-Air. Conservationist and journalist Paul Spencer Sochaczewski's book promotes Indonesia's biodiversity by vividly describing the achievements of Wallace and his assistant Ali, a young man who came from Borneo and who 'retired' in Ternate."

— *Pustanto, acting director of National Museum, Indonesia*

"Paul Spencer Sochaczewski has been a good friend over the years, not just to me personally but to the people of the province of North Maluku who are working to save our precious natural heritage. His latest book, *Look Here, Sir, What a Curious Bird*, reveals how important Wallace's discoveries were in our understanding of our region's biodiversity. The book also provides a detailed, touching, and insightful portrait of Ali, the young man who helped Wallace and who settled in Ternate after Wallace left for England. It's essential reading for anyone planning to visit this beautiful and historic part of Indonesia."

— *Azis Momanda, tourist official, North Maluku Tourist Office, and freelance guide specializing in Wallace-related travel*

"A fine work. Paul Spencer Sochaczewski has achieved the improbable — written a compelling 'enhanced biography' of an unheralded, little-known 19th-century teenager. Focusing on Ali, the Borneo-born assistant of Alfred Russel Wallace — the man who developed the theory of evolution before Darwin — Sochaczewski reveals how Wallace would not have been as successful as he was without Ali's support. Recounting Ali's adventures eloquently while exploring his relationship with Wallace, Sochaczewski is incisive, funny, and highly informative."

— *Peter Mikelbank, co-author of* Albert II de Monaco, l'homme et le prince

"Paul Spencer Sochaczewski has done a great service in bringing Ali, Wallace's faithful companion and camp manager, alive to us through this intriguing book. It could be argued that, without Ali, Wallace's achievements in the Malay Archipelago would have been much less successful. Among other interventions, Ali nursed Wallace during his famous malaria fit during which he had his epiphany of the Theory of Evolution by Natural Selection. I fully support the proposal that a replica of Wallace's house be erected in Ternate and that this education center should also commemorate the important role that Ali played in Wallace's achievements."

— *Nicholas Hughes, co-author of* The Quest for the Legendary House of Alfred Russel Wallace in Ternate

"Paul Spencer Sochaczewski's strength is that his mind, like that of Alfred Russel Wallace, seeks truth and accuracy but also relishes a broad palette of ideas, themes, and imaginings.

"In this easy-to-read volume, he offers two memorable heroes: Alfred Russel Wallace, a 19th-century giant, one of the world's greatest naturalists and collectors, and his 'faithful companion' Ali, who, up to now, has largely been overlooked.

"Sochaczewski, who has documented Wallace's life for more than 50 years, now digs into Ali's story and constructs a fascinating portrait of a young man who helped Wallace become as successful as he was.

"This intriguing book ranges far and wide and employs conventional evidence-based scholarship but also explores the realm of spiritualism — in the unusual but captivating guise of an open-minded non-believer who contacts mediums worldwide in an unrequited quest to uncover fresh information linked to the tantalizing mysteries surrounding Ali's life and legacy."

— *Derek Partridge, author of* What Makes You Clever: The Puzzle of Intelligence *and* Charles Darwin and a Tale of Two Theories *(in preparation), which reinterprets aspects of Darwin's development of the Theory of Evolution by Natural Selection*

"Not everything can be explained by science. Not everything can be explained by history. Paul Spencer Sochaczewski's literal and literary book about the search for Ali is a wonderful contribution to the multi-dimensional story of Alfred Russel Wallace. The wide-ranging themes covered by this book include scientific breakthroughs, the re-examining of what we know from history (and why history can be slippery), and the possibility that answers for some of the questions we pose may come from supernatural dimensions. Wallace was a philosophical naturalist, and I think he would have enjoyed the imaginative and speculative passages and conversations in the book, perhaps especially those involving mediums. I am certain he would have applauded Sochaczewski for giving Ali a voice.

"But the real enjoyment of reading *Look Here, Sir, What a Curious Bird*, came from the journey of investigation through which Sochaczewski leads us — as if we are co-conspirators in a detective story. I am reminded of J.B. Priestley's play *An Inspector Calls*. As the mysterious Inspector Goole says at one point: "One line of enquiry at a time." Sochaczewski follows the Ali trail systematically and engagingly. Do we find Ali in the end? Just as in the play, we may never know the answers by relying solely on the scientific method.

"The book commemorates Ali and Wallace, acknowledges

the role of Wallace's many assistants whilst travelling the Malay Archipelago, and helps to rebalance the history of one of the greatest adventures ever told."

— *Barry Clarke, organizer of the campaign to create a statue of Wallace and Ali in Singapore*

"Alfred Russel Wallace is best known as a ground-breaking naturalist, but he also played an indirect but important role in promoting the mid-19th-century exotic bird trade between eastern Indonesia and Europe. This book offers an insightful look at the relationship between Wallace and his able assistant Ali, who likely collected some 5,000 of the more thanw 8,050 rare birds that Wallace is given credit for."

— *Marc Argeloo, author of* Natuuramnesie *and historian focusing on the Ternate bird trade*

"Through the ages, our famous explorers and adventurers have rarely pursued their quests alone. They've been helped along by others who, barely acknowledged, have helped them on their way through seemingly impenetrable forests, swamps, and ravines. This was the case with Alfred Russel Wallace, the great Victorian naturalist whose daring expeditions were aided by a young man referred to simply as 'Ali.'

"This is an important, enlightening book not only because it tells Ali's story, at last giving him the prominence he deserves, but because it reminds us that we each have Alis in our lives, people to whom we should from time to time raise a grateful glass of thanks.

"Paul Spencer Sochaczewski, a conscientious scholar and gifted raconteur (a rare and welcome combination), pursues the unsung Wallace-Ali relationship (reminiscent of the classic Holmes-Watson partnership) through the distant corners of the Malay Archipelago and into the spiritual realm."

— *Indraneil Das, professor, Institute of Biodiversity and Environmental Conservation, University of Malaysia, Sarawak*

Also by
Paul Spencer Sochaczewski

Fiction

Redheads

EarthLove

Exceptional Encounters

Non-Fiction

A Conservation Notebook

Searching for Ganesha

Dead, But Still Kicking

An Inordinate Fondness for Beetles

Share Your Journey

Distant Greens

The Sultan and the Mermaid Queen

Malaysia: Heart of Southeast Asia

The Five-Book Non-Fiction Series

Curious Encounters of the Human Kind

Myanmar (Burma)

Indonesia

Himalaya

Borneo

Southeast Asia

Co-authored with Jeffrey McNeely

Soul of the Tiger

Eco-Bluff Your Way to Instant Environmental Credibility

Illustrations of the Postcard Maps of Southeast Asia on pages 55, 86, 98, and 126, by pomogayev/Adobe Stock.

Published by:
Explorer's Eye Press
Geneva, Switzerland

Editor and Project Manager: Marla Markman, MarlaMarkman.com
Cover and Interior Design: Kelly Cleary, kellymaureencleary@gmail.com

Publisher's Cataloging-in-Publication data
Names: Sochaczewski, Paul Spencer, author.
Title: "Look here , sir , what a curious bird " : searching for Ali , Alfred Russel Wallace's faithful companion / Paul Spencer Sochaczewski.
Description: Includes bibliographical references. | Geneva, Switzerland: Explorer's Eye Press, 2023.
Identifiers: ISBN: 978-2-940573-41-7 (paperback) | 978-2-940573-42-4 (epub)
Subjects: LCSH Wallace, Alfred Russel, 1823-1913. | Wallace, Alfred Russel, 1823-1913--Friends and associates. | Wallace, Alfred Russel, 1823-1913--Travel--Malay Archipelago. | Wallace, Alfred Russel, 1823-1913--Travel--Indonesia. | Naturalists--England--Biography. | Science--Great Britain--History--19th century. | BISAC BIOGRAPHY & AUTOBIOGRAPHY / Adventurers & Explorers | BIOGRAPHY & AUTOBIOGRAPHY / Environmentalists & Naturalists
Classification: LCC QH31.W2 .S63 2023 | DDC 508/.092--dc23

978-2-940573-41-7 (Print)
978-2-940573-42-4 (E-Readers)

Printed in the United States of America

DEDICATION

It seems as if Ali is destined to remain an enigma, a lost soul in the large filing cabinet labeled: People Who Did Important Things But Never Got Sufficient Credit.

I suspect we all have our share of Alis in our memory banks and karmic stockpiles, folks who have helped us in innumerable ways but to whom we rarely raise a glass. We don't know Ali's birthday. But perhaps we should declare January 8 — Alfred Russel Wallace's birthday — as Global Ali Day, to give remembrance to those who nursed us, granted us unexpected favors, and who, sometimes without our awareness, eased our paths.

Contents

Author's Note

Searching for Ali is one of my long-term quests.

I seek tiger magicians in Sumatra, Snowmen of the Jungle in Flores, the mythical mountain where Hindu monkey god Hanuman collected medicinal plants, and the location of "Waltzing Banana Island." I want to know why people believe in the power of magic amulets and why sultans consort with mermaids. I'm fascinated by white elephants. I explore the secrets of growing jumbo, juicy tomatoes. I search for a writing voice that makes people sit up and take notice. I don't expect readers to leap and shout "huzzah!" but a satisfied smile once in a while would be welcome. I would like to get into the head of a medium. I would like to meet someone who can explain the Higgs boson.

And I have spent more than 50 years trying to find out more about Ali: who he was, why he went with Wallace, what his contribution was, where he ultimately settled, and whether I can contact his descendants.

During this journey I've come to appreciate how difficult it is to write history — how truth is an elusive concept, not just when writing about people long dead but with our own personal stories as well. And it's helped me recognize that my life — and yours, I bet — has been shaped by many Alis who have eased my journey, opened doors, and helped me along the way, often without adequate recognition or thanks.

Ali was then, and remains, a footnote, an historical afterthought written in small type. Does he deserve better?

SECTION I

In Search of Ali

CHAPTER 1

The Hero's Journey

Consider a journey. Most journeys, both the physical and the psychological, involve movement—a goal, an adventure, a return.

Alfred Russel Wallace's great life journey got a kickstart when he met Henry Walter Bates in England, in 1844, when Wallace was teaching at the Leicester Collegiate School. Like Wallace, who left school at the age of 14, Bates also left school at an early age, 13, and went to work as an apprentice stocking-maker in the family business. They were both keen amateur entomologists and together explored the English countryside in search of beetles and butterflies. In one of the great examples of youthful, impetuous impulses, they said: *Hey, let's go to Brazil.* Wallace was 25, Bates, 23. Neither had ever left Britain. They also had little money and no diplomatic support. Neither spoke Portuguese. A pessimistic observer would say all the stars were aligned for failure.

They engaged Samuel Stevens in London to act as their agent to sell the specimens they collected to European collectors. They compiled "want lists" from museums and collectors and visited London's museums as part of a crash course on Brazilian biology and culture.

The trip turned out to be a turning point for both.

At first they travelled together but separated when Bates remained on the Amazon River while Wallace left to explore the Rio Negro.[1]

The classic hero's journey (more on this later) is punctuated by setbacks. Wallace experienced several major tragedies by the age of 29.

After travelling to Brazil to work with Alfred as a collector, Wallace's younger brother Herbert died of yellow fever, an event that caused great grief (and one might imagine, guilt) in the young Wallace.

Wallace himself was often ill or injured; he caught malaria and suffered poor health that would follow him to Asia.

But it was the next traumatic experience that is the stuff of high drama.

After four years Wallace had had enough of the tropics. Upon returning downriver from the upper Rio Negro, he found, to his frustration, that most of his collection, including numerous specimens that were new to science, had not been shipped as he had instructed. These crates contained:

> a considerable collection of birds, insects, reptiles and fishes ... consisting of about twenty cases and packages. Nearly half of these had been left by me at Barra a year before to be sent home; but a new government, arriving there shortly after I left, took it into their heads that I was engaged in a contraband trade, and so I found them still there on my way down, in the present year, and had to bring them all with me.[2]

Burdened by his treasures, he made his way downriver to the port of Para (now Belém) and booked passage on a rusty brig christened the *Helen*.

Wallace was far from being in robust health, writing that he was "suffering from fever and ague, which had nearly killed me ten months before on the upper Rio Negro, and from which I had never since been free."

At 9 a.m. on August 6, some 700 miles east of Bermuda, the *Helen* caught fire, and all aboard hurriedly abandoned ship:

> The only things which I saved were my watch, my drawings of fishes, and a portion of my notes and journals. Most of my journals, notes on the habits of animals, and drawings of the transformations of insects, were lost.[3]

Most retellings of this adventure incorrectly report that with the sinking of the *Helen*, Wallace lost *all* of his South American collection. This is an example of how accuracy suffers when one dramatic claim ("he lost his *entire* collection") is put forth, then repeated countless times in identical narratives. To set the record straight, Wallace did *not* lose his entire collection. He had previously successfully shipped Stevens some 27 bird skins from the Amazon; these are now in the collection of the Natural History Museum, UK.

The loss of most of his collection as a result of the *Helen* tragedy would have been enough to break a lesser man. Imagine four years of hard labor, visiting places no naturalist had explored before, suffering from loneliness and jungle fevers, and having to rely on his wits and the kindness of strangers while collecting hundreds of specimens that almost certainly were new to science, and therefore of great interest to British collectors and the scientific establishment:

> My collections were mostly from the country about the sources of the Rio Negro and Orinooko [Orinoco], one of the wildest and least known parts of South America ... I had a fine collection of the river tortoises (Chelydidae) consisting of ten species, many of which I believe were new. Also upwards of a hundred species of the little known fishes of the Rio Negro ... My private collection of Lepidoptera ... there must have been at least a hundred new and unique species. I also had a number of curious Coleoptera, several species of ants in all their different states, and complete skeletons and skins of an ant-eater and cow-fish, *(Manatus)*; the

> whole of which, together with a small collection of living monkeys, parrots, macaws, and other birds, are irrecoverably lost.[4]

The lifeboat of the *Helen* leaked, the winds were against Wallace and the crew, there was no protection from the sun and rain, and food and water were in short supply. After 10 days and nights, he writes: "We heard the joyful cry of 'Sail ho!' and a few hours hard rowing got on board the 'Jordeson' from Cuba, bound for London."

They were off the lifeboat and alive, but life aboard the *Jordeson* was difficult:

> [We] soon got to be very short of provisions... had not two vessels assisted us with provisions at different times, we should actually have starved; and as it was, for a considerable time we had nothing but biscuit and water. We encountered three very heavy gales, which split and carried away some of the strongest sails in the ship, and made her leak so much that the pumps could with difficulty keep her free.[5]

But Wallace returned to England, weakened and no doubt traumatized. He swore never to go to sea again.

He had arrived at one of the decision points common to all heroes' journeys. Some men, after surviving what Wallace had gone through, would have chosen to settle far from the coast and work in an accounting office and stay home to grow roses. Some men would have been so devastated they would have turned to alcohol or drugs or have been haunted by depression and nightmares. Wallace, however, realized that he had to get back on the horse.

He realized that to be taken seriously by the scientific community he had to produce academic papers based on a new collection, from a region that had not been intensively studied. He also had to earn a living, which meant going to one of the most biologically diverse areas on the planet. He applied to the Royal Geographical Society for a grant to travel to Southeast Asia and sailed to Singapore from Southampton, first aboard the P&O steamer *Euxine*, then the *Bengal*, and finally on the *Pottinger*. The journey took six weeks.

He spent the first months in Singapore getting acclimatized and collecting a significant number of beetles around Bukit Timah, Singapore's highest peak at 538 feet.

While in Singapore, sometime during the period from September to October 1854, Wallace visited James Brooke, the famed "White Rajah" of Sarawak, who invited him to Borneo.

Wallace's prior four-year Amazon adventure had helped him immensely in learning to survive in alien environments and cultures. But in Asia he was a stranger in a strange land. Nevertheless, he was a quick learner and rapidly gained competence in Malay, the lingua franca of the region.

Like all good journeys, Wallace's voyage was filled with challenges. He was travelling independently – he had no government or military support system. He also had little cash – he earned enough to survive by sending natural history specimens to Samuel Stevens, his agent in London, who then sold the critters to enthusiastic collectors. (Darwin, on the other hand, during his famous voyage on the *HMS Beagle*, lived onboard in what was, in effect, a floating base camp, with Royal Navy sailors on hand to provide security, logistics, laundry, and food. He had a permanent, dry place to write his notes and mount his specimens. The downside: Darwin shared a cabin with the captain, Robert FitzRoy, who, Darwin noted, had a quick temper that resulted in behavior sometimes "bordering on insanity.")

Just as Wallace learned and evolved, Ali was on his own journey of discovery.

Starting out as a cook, Ali himself grew and became a valuable assistant. He learned to collect and mount specimens. He took on more responsibility for organizing travel (just imagine the negotiations with self-important village chiefs, unreliable porters and laborers, and greedy merchants, whose eyes no doubt grew large when they saw a white man like Alfred come to buy supplies). Ali was a servant, but the pair developed a friendship; Wallace called him "my faithful companion."

If it hadn't been for Wallace's writings, Ali would have remained one of millions of, what historian John Miksic calls, "people without history."

There's an awful lot we don't know about Ali. We speculate on where he came from. We aren't sure how Wallace met him, whether he ever went to school, or where he settled after he parted ways with Wallace. And most interesting (and frustrating), we have no idea what Ali thought of the tall, awkward, bearded Englishman who spent his days collecting innumerable insects and his nights writing in a small notebook by the dim light of an oil lamp.

Colonial history has been generally written by energetic, spirited, over-dressed, well-educated Portuguese, Spanish, Dutch, German, British, and French explorers who came in the service of god, royalty, science, and greed. Can we speculate how their stories would have been recounted by the people who received the foreign visitors? Ali's story has been told by Alfred Russel Wallace. But what would Ali have written about their relationship?

I wonder what we might learn by flipping things around and looking at the Wallace-Ali relationship from Ali's point of view. What did Ali think about all those characteristics (well-educated, science-influenced, relatively wealthy, curious, much-travelled, open-minded, meticulous, living far from his family, willing to sacrifice comfort for the sake of the quest) that intrigues a modern audience about Wallace? Can we speculate how Ali judged this tall, gawky, bearded (beards are a constant source of fascination, and often fear, to many rural Asians) employer? Did Ali defend Wallace when villagers thought he was an evil demon?[6] At times Ali might have been confused. Wallace was his boss, his superior, his colonial master, but he cared for Ali when he was sick, almost as a father might have done. This wasn't the way an Englishman was expected to treat a native servant. And the other Englishmen Ali had seen were rich, while Wallace had to sell insects and birds to earn enough money to pay Ali's meager salary. Ali's contemporaries had tuans who were Big Men, but Wallace was the white man's equivalent of a kampong lad — no military medals, or grants from the queen, or noble title. How did Ali try to make sense of this foreigner, who, with all the wealth and status that designation implies, deliberately chose not to live like other white foreigners but instead insisted on spending challenging weeks camped out in the forest, in palm-leaf shelters that leaked, fighting rats, dogs, and ants that tried to devour his specimens. And my word, those specimens. Often, they were dull in color and to Ali without interest. Yet Wallace seemed as happy to collect an unremarkable-appearing grey-brown beetle as tiny as a fly with the same enthusiasm as he collected a big, bold hornbill. And what mental illness drove Wallace to spend long periods huddled over his journal writing intently about an ant. A miserable *semut*! And so many of them! Wallace said they were all different, but Ali couldn't see much beyond the fact that some were big and some were small and some were black and some were red. Wallace ate barely palatable food — pythons; small birds; small, bony fish — just like the laborers (although he grumbled about it). He actually enjoyed durian. And, we might speculate, Wallace smelled. Even in the forest, Ali took pride in his hygiene, washing his clothes and himself at every opportunity with buckets of rainwater or in a nearby stream. But, he had to admit, some of the butterflies and birds *were* attractive. However, all told, if Ali was in control of the shotguns he might better have used his time and ability to shoot deer for the dinner table.

I coach writers, and I particularly enjoy working with authors who are writing their personal stories. One of the first, and among the most important, concepts I encourage writers to understand is the "hero's journey."

Christopher Vogler is a Hollywood screenwriter whose book *The Writer's Journey* explores how a classical mythic structure was used in iconic films such as *Casablanca*, *The Wizard of Oz*, *Star Wars*, and *Close Encounters of the Third Kind*. "All stories consist of a few common structural elements found universally in myths, fairy tales, dreams, and movies," Vogler says. "They are known collectively as The Hero's Journey."

One essential archetype in the hero's journey is the character Vogler calls the Mentor (and which philosopher Joseph Campbell, author of *The Hero with a Thousand Faces*, refers to as the Wise Old Man or the Wise Old Woman), who teaches and protects heroes. In classical mythology, as well as contemporary novels, the Mentor is a sage adult — Merlin guiding King Arthur, the Fairy Godmother helping Cinderella, or a veteran sergeant giving advice to a rookie cop.[7]

If you wish to plot Ali's hero's journey in this way, then Wallace is clearly the Mentor, and Ali takes the role of a keen, but initially naïve, son who grows in competence and confidence as the story progresses. Wallace took him on a magnificent journey, and by the time Ali and Wallace parted company, Ali no doubt had matured considerably.

In addition to Campbell's archetypes, I suggest there is another key element to a hero's journey, whether fictional or real. That is the "decision point," the moment when, often following a crisis, the hero is confronted by a major choice, a crossroad, a life redirection, a safe or a risky option. Choose one path and your life changes in a certain way, choose another and you veer off into an alternate reality. Wallace had many such critical decision points – when he decided to go to the Amazon, when he chose to book passage aboard the *Helen* for himself and his collection, and when he decided to overcome the trauma and heartbreak of the sinking of the *Helen* and continue exploring.

Ali, likewise, had a decision to make, although he might not have considered it as such. He chose to work for Wallace. The result? As Jerry Drawhorn, of Sacramento State University, notes: "Ali would have been one of the most widely travelled Malays of his age. [With Wallace] he would have seen most of what is today modern Indonesia. He would have seen the ancient Hindu temples of Java; the modern metropolises of Batavia and Singapore; the primitive villages of the people of Dorey and the stylized royal courts of central Java. He tasted modern science and medicine and yet retained his beliefs in ghosts and men who could transform into tigers."[8]

The reverse is also true.

Just as Wallace taught and guided Ali through new adventures, perspectives and skills, Ali was similarly Wallace's guide. The Englishman was a good pupil, and Wallace's steep learning curve, abetted by Ali, included mastering the Malay language, understanding the vagaries of dozens of cultural groups (Wallace compiled 57 vocabularies of indigenous groups he visited during his travels in Asia) and becoming comfortable with an Asian worldview in which female man-hating ghosts, crocodile whisperers, bird omens, and kings who consort with mermaids, danced an intricate waltz of life as seen through a rice-wine-infused spectral prism.

Wallace's story is entwined with Ali's. The challenge (and fun) lies in trying to document that joint venture.

CHAPTER 2

Slippery History, Sloppy Memory, Creative Conjecture

What did you have for lunch last Thursday?

Our memories are elusive creatures. So is history.

If I were to write about an adventure I had a few months ago, the information I give you would, of necessity, but also by choice, be filtered through the leaky webbing of my memory, by the censor on my shoulder, by the flow and vocabulary of my prose or style of speaking, by my connection with the person receiving my tale.

Can you remember, verbatim, word for word, a conversation you had an hour ago?

Can you recall an event without amplifying, pruning, enhancing?

It's impossible.

Most everything we know about the young man named Ali comes from Alfred Russel Wallace's writing. It is information that might well be accurate (for Wallace was a careful writer), but nonetheless it is filtered information that has been processed through Wallace's internal editor.

When we speak or write, we prefer to lock our skeletons in a dreary cellar and encourage beautiful memory-angels to frolic in the light.

We tell a story differently to each listener. And each listener hears that story selectively. Different people tell stories about the same incident with their own particular spin. We live in a *Rashomon*-like world of ambiguity.

We aren't always aware of the subtleties and biases of others (we barely understand our own inconsistencies). A mafia boss might appear evil, but she might adore her children. A primary school teacher might be kind to the children under his care but harbor thoughts about dismembering his partner. The only reliable truth is that our "true" stories are imbued with a wallop of fiction – dramatic recreations based on elusive and selected ideas that we present to our friends as facts.

"HOW DO YOU EXPLAIN TODAY'S LESSON WAS DIFFERENT THAN WHAT'S ON WIKIPEDIA?"

CartoonStock.com

With so many "truths" floating around, it's sometimes hard to decide who to believe.

Broadly speaking, there are two species of historians. Those who seek broad truths and those who seek specific facts.

The broad-truth seeker tries to make sense out of the big picture – the events, the underlying dynamics, and the psychology of the players.

The second form might be termed the "detective" – scholars who embark on Sherlock Holmes-like crusades to sift through dusty clues in search of an "aha" moment they can use to write what they consider to be an accurate account of an event or life.

It's all a bit as Hercule Poirot said when he compared the work of the archaeologist (think of Jean-François Champollion deciphering the Rosetta Stone) with his own detective work: "You take away the loose earth, and you scrape here and there with a knife until finally your object is there, all alone, ready to be drawn and photographed with no extraneous matter confusing it. That is what I have been seeking to do — clear away the extraneous matter so that we can see the truth — the naked shining truth."

Of course, life is seldom as neatly binary as Agatha Christie might have us believe. Daniel Warner, author of *An Ethic of Responsibility in International Relations*, says: "In writing history there is no absolute 'Truth,' only a multitude of lesser lower-case facsimiles we might term 'truths.'"

When writing about historical characters, historians search for the holy grail that "proves" a speculation.

In the hierarchy of research, the most sought-after grails would be the letters, journals, and autobiographies written by the subject herself.

Next would be comments written by a contemporary of the subject, the situation in which Wallace writes about Ali.

Third would be a trove of academic, peer-reviewed, multi-footnoted journal articles and books by respected researchers, many of which are cited in this volume.

From these sources, historians then evaluate and triangulate until, like Miss Marple, they reach a satisfactory solution to the puzzle. (Or as cynical Napoleon said: "History is a set of lies that people have agreed upon.") This conclusion would then become the generally accepted "fact" and enshrined as conventional wisdom, at least until a similarly determined researcher comes along, unearths new clues, and "disproves" the earlier historian's theory.[9]

When writing about Ali, all we have to go on are the narratives written by Alfred Russel Wallace, and, to a lesser degree, by Spenser St. John, Frederick Boyle, and Thomas Barbour. Wallace was a scientist, who was mostly meticulous in his record-keeping — he left scores of notebooks, inventories, letters, and journals. Luckily, he was also a prolific diarist and recounted episodes (which often mentioned Ali) during his eight years in Southeast Asia. Nevertheless, he is just one source through which we learn about Ali's story, and he, like every writer everywhere, wrote through his individual prism.

The result? An historian can work like a code-chasing hero in a Dan Brown adventure, or like Alan Turing, breaking WWII German ciphers. It's an academic treasure hunt, or, on a more primal level, a hunter tracking a deer or an undercover cop infiltrating a mob. The quest is scintillating, and the hunt is manifested by a desktop submerged in photocopies, notes, and books and a computer overflowing with Google bookmarks. The hope is that the effort will result in a prize, an answer, an insight — at the very least, a good guess. While we love mysteries, we dislike unreconciled cliff-hangers. We want mysteries to be *solved*. Yet sometimes, as in this book about Ali, there are few clear answers. We have to make do with educated speculation, and take the liberty of venturing a guess on what *might* have happened, what he *might* have said, what he *might* have thought. And why not sprinkle a few just-over-the-top scenarios, simply to see if anyone's paying attention? Let's call this approach an "enhanced biography."

Let us not ignore the psycho-historians, who search for the deepest, most personal, often-embarrassing, sometimes traumatic, hopefully scandalous, life-changing incidents that have shaped a person's personality. These are the people who realize that history is more complex than a recital of empirical facts. They hope to discover a scandal, a suicide, an injustice, a Big Lie, a conspiracy, an act of greed, or an act of god that will explain a subject's curious behavior (think of the pivotal incident that formed the behavior of Lord Jim in Joseph Conrad's novel). Psycho-historians are people who like nothing better than to rip away the curtain hiding the venerated Wizard of Oz to reveal a bombastic, short, fat, arrogant man with a Napoleon complex, who, no doubt, they might postulate, had been terrified of his father, over-cuddled by his mother, and summarily ignored by women he lusted after. They might stick, poke, and speculate like a psychiatrist, but whereas the professional therapist is bound by a code of confidentiality, the psycho-historian feels no hesitation in exposing the subject's naked psyche. Just the opposite: The warts, peccadillos, angst, and never-to-be-fulfilled dreams of the subject *must* be exposed for the greater glory of *history*.

What psychological triggers might have catalyzed Wallace's desire to leave England and explore the world? How did he overcome the trauma of losing much of his South American collection, and almost his life, when the ship he was on caught fire and sank? How did he manage to compartmentalize so successfully? In *The Malay Archipelago* he writes about shooting and killing a female orangutan, rescuing the baby she had been nursing, trying to care for the infant, writing sentimental paragraphs about how cute it was, and then, when the baby died, boiling it and selling the skeleton.

And what went on in Ali's mind? Of course, there is no written record of his background or his thoughts. Why did he leave home to travel with a foreigner? What was the trigger to his wanderlust – was it a cultural trait, a personal imperative, or just a way to earn money? Why was he so loyal? Did he have ambitions of his own, or was he happy just to go along for the ride? Did he think Wallace was an entitled but silly white man who used perfectly good guns to shoot tiny birds for his collection instead of deer for a barbecue? And what about all those too-tall, too-loud, over-dressed white men whom Wallace called friends, with their impressive technology, huge houses, and seemingly innate ability to read, write, and apparently rule the world? Were they cash cows waiting to be milked? Were they to be respected, admired, even emulated? Or were they foreign interlopers, temporary irritants in Allah's grand scheme of things?

"I know what you're thinking, but let me offer a competing narrative."
CartoonStock.com

For every incident, there is generally an alternative explanation.

Listen to a conversation. We rarely speak in complete, coherent sentences. We speak in lightning-flash phrases, accented by body language, adjusted to meet the interest and intellectual compatibility of the listener.

We complain and praise. We spew unwarranted slander and unrequited adulation. We grumble, embellish, exaggerate, and utter frothy nonsense. We misremember, sometimes intentionally. We tell harmless white lies to save someone's feelings, and we spew grand Donald Trump-sized lies just for the hell of it. We write true-appearing stories spiced with irony, subtlety, and "wink-wink" clues that sometimes are interpreted as gospel (see Isabelle Desjeux in Chapter 8, "The Pain of Changing an Idea Thought to Be True"). We have unending opinions and points of view. The stories we tell – "news" that we got from a website, and gossip, and rumors of all sorts – are as changeable as Scottish weather and as unreliable as a politician's promise. We live in a world of continual Chinese whispers.

Few people are capable, or willing to follow the admonition of Sergeant Joe Friday in the 1950s TV series *Dragnet*: "All we want are the facts, ma'am."[10]

Trustees of the Natural History Museum, London

Alfred Russel Wallace kept detailed journals of his daily activities and notebooks (two of the dozens are shown here) of the "natural productions" he collected. He maintained this habit even while living in difficult conditions, perhaps in a hut that might have leaked and been cold, without a proper desk or lantern. He was a thorough chronicler of his life, and these journals gave him the material to write *The Malay Archipelago*, published in 1869, some seven years after he returned to England. Scholars have pointed out several places where he "misremembers," but overall we can trust his writing, which, of course, is the primary documentation we have about Ali.

When writing about a modern personage, there is often an abundance of contemporary accounts, letters, and, in some cases, first-person witnesses. But with historical characters, historians are forced to rely on chroniclers, who vary widely in their veracity.

This short book relies, of necessity, on Wallace's memory and selective thoughts. Are his written statements neutral? Of course not. Conversely, the same phenomenon occurs when you hear something. We listen selectively; reality is in the ear of the beholder.[11]

A final thought about writing history. The historian is, by definition and I suspect by nature, a left-brained, rational, logical, Cartesian-oriented person who prefers to see the world in terms of black and white, truth or fiction, verifiable evidence or imprecise conjecture. But one thing that living in Asia for many years has taught me is that it can be dispiriting to try to over-rationalize one's emotions, beliefs, and behaviors. There are certain life concepts that should not be dissected and put under a microscope, for, at the end of the day, to paraphrase Gertrude Stein, "often there is no there there." Alfred Russel Wallace believed in spirits and tried to explain his certainty by using scientific analysis. That's a direction that leads to confusion, or, as the British say, "getting your knickers in a twist." If you want to believe in spirits, or religion, love, hate, or beauty, then go ahead and do so. These things are the basic ingredients of human life — logic does not conquer emotion. As Sultan Hamengkubuwono IX of Yogyakarta told me when I asked about his relationship with the Mermaid Queen, a spirit named Kanjeng Ratu Kidul who purportedly became consort of Prince Senopati, the founder of the dynasty, and who promised to support his royal descendants: "Part of my brain is a logical Dutch-educated brain. Part of my brain is an emotional Javanese brain. The Dutch ask questions and try to analyze every action. The Javanese accept that not all things can be dissected into their component parts. So, don't ask so many questions that demand a yes or no answer. There are some things you can't explain. Simply believe it. Or not."

This book goes in a direction that might upset historians who base their life work on what has been *written*, often by other experts. I have included detailed references from published sources but have also taken the liberty of speculating, which is a wonderfully liberating activity. I might well be wrong; trying to unravel the subtleties and intentions of all relationships (even our own) is impossible. All relationships contain a large smidgen of we'll-never-know details and never-to-be-exposed deceit. I don't know what really was going through the minds of Wallace and Ali, but that won't stop me from venturing a guess. We all make our own stories, and we attempt to fill in the blanks in the stories of others.[12 13]

SECTION II

Wallace and Ali —
A Curious,
Productive Partnership

CHAPTER 3

Everybody Needs Somebody

Western history lionizes famous explorers as larger-than-life individuals who braved the elements alone – stoic, unflappable, and with immense strength of character and fortitude. But actually, most explorers, the famous as well as the overlooked, the real and the fictional, relied on often-unheralded people to assist in their odyssey. Just as Hindus implore Ganesha, the elephant-headed god, to remove obstacles, Lewis and Clark relied on Sacagawea to lead the way and make their journey successful. Don Quixote teamed with Sancho Panza just as Sherlock Holmes consulted Dr. John Watson, and Don Giovanni needed Leporello to manage the accounts of his conquests. Batman was supported by Robin, and Dante was inspired by Beatrice Portinari. Mountaineer Edmund Hillary climbed Mount Everest with Tenzing Norgay (but in fairness, Hillary gave Tenzing Norgay full credit when they summited in 1952 and refused to identify which of them first stood on the top). Ferdinand Magellan counted on Enrique, a slave he bought in Malacca and whom he "encouraged" to renounce Islam and convert to Catholicism. Enrique became Magellan's interpreter, navigator, guide, and assistant; he may also have been among the first people to complete the circumnavigation of the globe, since Magellan, who gets the credit, was killed in the Philippines in 1521 before finishing the epic journey.[14]

British naturalist and explorer Alfred Russel Wallace was supported by a teenager we know by one name, Ali, and without Ali's assistance, it is unlikely Wallace would have been as successful as he was.

Wallace hired Ali in 1855 in Sarawak, now a Malaysian state on the island of Borneo.

Wallace wrote:

> When I was at Sarawak in 1855 I engaged a Malay boy named Ali as a personal servant, and also to help me to learn the Malay language by the necessity of constant communication with him. He was attentive and clean, and could cook very well. He soon learnt to shoot birds, to skin them properly, and latterly even to put up the skins very neatly. Of course he was a good boatman, as are all Malays, and in all the difficulties or dangers of our journeys he was quite undisturbed and ready to do anything required of him.[15]

A simple statement of fact that opens the door for considerable speculation.

For a start, under what circumstances did Wallace hire Ali?

What was their relationship?

What happened to Ali after Wallace left Asia to return to England?

Why should we care about the curious partnership between Wallace and Ali? Is there resonance with our own lives?

One measure of Ali's importance is that in *The Malay Archipelago* Wallace mentions Ali by name no fewer than 42 times.

Wallace lived by collecting "natural productions," and selling them via Samuel Stevens, his "beetle agent" in London. Wallace was smart (and humble) enough to realize he needed help to collect these curiosities.

Historian John van Wyhe estimates that "at a minimum, 1,300 people, mostly residents of what is now Indonesia, helped Wallace achieve his great work of natural history, and given the incompleteness of the written record the true number could easily be more than twice this number."[16]

Of course, the "veritable army" that van Wyhe records includes helpers of all sorts – boatmen, carpenters, camp workers, wildlife spotters, as well as collecting assistants.

Some were unskilled laborers who (generally) did what they were asked. Some workers had specific, one-off tasks. When an orangutan Wallace had shot in Sarawak remained stuck in the upper branches of a tall tree, he asked "two Chinamen with axes to cut down the tree," and on another similar occasion, he wrote: "Two Malays, on the offer of a dollar, climbed the tree, and let down the dried remains [of a dead orangutan]."[17]

Similarly, in Palembang, Sumatra, when a female hornbill was discovered in a nest in a tree, Wallace "offered a rupee to any one who would go up and get out the bird." Eventually the bird and its chick were brought to him, and these were illustrated with a woodcut in *The Malay Archipelago*."[18]

Wallace's collecting assistants numbered more than 30 who worked full time; they displayed considerable initiative, a feature encouraged by Wallace. Van Wyhe writes: "Wallace's assistants were certainly not passive collecting drones. They were often intelligent young men who used their experience and wits to find new sources for the things that so interested their enthusiastic employer."

MY HOUSE AT BESSIR, IN WAIGIOU.

Alfred Russel Wallace, The Malay Archipelago

Alfred Russel Wallace's simple house in Waigiou (now Waigeo), an island in the far-eastern corner of Indonesia in the province of West Papua, part of the Indonesian half of the island of New Guinea.

Wallace required up to 13 men to help him move and set up camp, a process he went through approximately a hundred times during his eight-year expedition in Southeast Asia. Once established, his living conditions were basic, and he relied on Ali and other members of his staff to maintain the encampment. He also often required the services of translators, since the people of some of the isolated islands he visited did not speak Malay, the lingua franca of the archipelago. I admire his perseverance and energy — how does one man, travelling alone, manage to collect, identify, and prepare 125,600 specimens of beetles, butterflies, birds, and mammals (including 17 orangutans)? He then had to store these specimens so that the ants, rats and dogs roaming the camp would not eat them. He was a remarkable taxonomist, particularly considering that he had no formal training in taxonomy or biology. He then had to pack these specimens so they would survive a multi-month journey to England. Such an achievement would tax even a modern naturalist with easy access to logistics and information resources.

With the help of Ali, and other individuals, Wallace's tally of beetles, ants, butterflies, orangutans, birds of all descriptions, and curiosities like gliding squirrels was, by any measure, remarkable. During his eight-year voyage throughout territories that are now the countries of Singapore, Malaysia, Indonesia, and Timor-Leste, he collected, by his own reckoning, 125,660 individual specimens, which consisted of 310 mammals, 100 reptiles, 8,050 birds (more on this later), 7,500 shells, 13,100 butterflies and moths, 83,200 beetles, and 13,400 other types of insects.

This haul included an astounding tally of creatures new to Western science: 50 *new* species of butterflies, 900 *new* species of beetles, 212 *new* species of birds, 50 *new* species of butterflies, and, perhaps the most remarkable, 200 *new* species of ants.

Wallace travelled with no official government support, no diplomatic credentials, no university grant, no military battleship waiting in the background to rescue him in case of trouble. He had to rely on his wits, personality, and initiative.

Wallace was a bundle of contradictions. He was proud of his logical thinking but wasn't afraid to show his sentimentality. He was a rational man who earnestly believed in spirits. He longed to settle down with a good wife but crowed about his peripatetic and exotic lifestyle. He complained about the hardships he endured, but he loved visiting out-of-the-way places. In 1860 he teased a friend in London: "I defy all the members of the Royal Geographical Society in full conclave to tell you where is the place from which I date this letter." He was in Bessir, a village on the island of Waigiou (now Besir on Waigeo) off the northwest tip of the island of New Guinea, then (and now) barely known outside the immediate vicinity.

He was a man of both big ideas and attention to detail. He was curious, and, when he chose to be, successfully social. To use modern terms, he learned to network with people in authority in order to obtain permission to travel and collect. Without much personal money to spend, he asked for, and generally received, the loans of houses and workers, letters of introduction, and various forms of support from Dutch officials, missionaries, and numerous sultans, rajahs, and local officials. The Earl of Cranbrook, a leading ornithologist and mammologist, and historian Adrian G. Marshall recognized that "his position was enhanced by the standing he gained by the recommendations of people in authority at all levels, from Governor to local ruler or head man."[19]

Nevertheless, there were many times when he couldn't rely on the kindness of strangers. In order to accomplish his goals, he had to hire, cajole, bribe, or threaten local assistants. Some of these assistants were collectors — local men whom he paid by the piece. Cranbrook and Marshall wrote that Wallace regularly "bought trade specimens [and] was always ready to recruit casual help in collecting, whether small boys with blowpipes or local hunters and bird traders."[20 21]

But several helpers were valuable assistants who were important enough to merit repeated mentions in his memoirs.

Alfred Russel Wallace, The Malay Archipelago. *Wallace's gliding frog, Rhacophorus nigropalmatus. Illustration by J. G. Keulemans based on a watercolor painted by Wallace.*

Some of Wallace's most interesting "natural productions" were collected by unidentified helpers. Wallace wrote: "One of the most curious and interesting reptiles which I met with in Borneo was a large tree-frog, which was brought me by one of the Chinese workmen. He assured me that he had seen it come down in a slanting direction from a high tree, as if it flew. On examining it, I found the toes very long and fully webbed to their very extremity, so that when expanded they offered a surface much larger than the body. ... This is, I believe, the first instance known of a 'flying frog,' and it is very interesting to Darwinians as showing that the variability of the toes which have been already modified for purposes of swimming and adhesive climbing, have been taken advantage of to enable an allied species to pass through the air like the flying lizard."[22]

Wallace wrote about 14-year-old Charles Martin Allen, who travelled with Wallace from Britain to Singapore and stayed with Wallace for about a year and a half.

Wallace's early opinion about the young man started with cautious optimism:

> Charles gets on pretty well in health, and catches a few insects, but he is very untidy, as you may imagine by his clothes being all torn to pieces by the time we arrived here. He will no doubt improve and will soon be useful.[23]

Things quickly unraveled:

> If it were not for the expense I would send Charles home; I think I could not have chanced upon a more untidy or careless boy... He is very strong & able to do any thing, but can be trusted to do nothing out of my sight.[24]

And Wallace got increasingly exasperated when Allen couldn't mount specimens properly:

> In every thing it is the same, what ought to be straight is always put crooked. This after 12 months constant practice & constant teaching... I believe he never will improve.[25]

And Wallace hoped that his next assistant would be more competent:

> Another with a similar incapacity would drive me mad.[26]

In spite of the personality differences (for Wallace himself was not the easiest person to work for), Allen helped Wallace collect nocturnal insects during a visit to Peninjauh (now Peninjau, near the town of Siniawan) in Sarawak. This expedition, incidentally, was where Wallace might have met Ali, who was likely travelling with Spenser St. John (more on this in Chapter 9, "Ali's Origin – Sarawak?"). Wallace recorded an extraordinary haul:

> On good nights I was able to capture from a hundred to two hundred and fifty moths, and these comprised on each occasion from half to two-thirds that number of distinct species ... on twenty-six nights I collected 1,386 moths ... during the six succeeding years, I was never once able to make any collections at all approaching those at Sarawak.[27]

In *The Malay Archipelago*, Wallace writes several times in the first person, taking full credit for the Peninjauh collection. But in his *Insect Notebook*, Wallace acknowledges that there was a division of labor. Drawhorn writes: "Wallace focused on collecting moths and Allen captured beetles, wasps, termites, and other insects . . . on two nights actually collecting more specimens than Wallace."

When Allen finally left him, to train as a teacher in the Anglican mission school in Kuching, Wallace wrote:

> I hardly know whether to be glad or sorry he has left. It saves me a great deal of trouble & annoyance & I feel it quite a relief to be without him.[28]

Yet some four years later, Allen rejoined Wallace, having matured into a reliable collector, notably of birds. Wallace sent him to collect independently, particularly in what is now Indonesian New Guinea. John van Wyhe, of the National University of Singapore, and Jerry Drawhorn, of Sacramento State University, have scoured Wallace's notebooks and journals and calculate that Allen and his team collected between 30,000 to 48,000 specimens, an astonishing one-quarter to one-third of Wallace's total haul. Among these treasures, they speculate "that Allen collected 2,900 of Wallace's total of 8,050 birds."[29] But Ali is likely to have even more birds to his credit (see Chapter 6, "Ali's Career Path – Bird Hunter, Taxidermist").

Then there is the up-and-down relationship with a young man named Baderoon, whom Wallace had hired in Celebes (now Sulawesi). Wallace had little time for people (especially boys on the payroll) who were lazy or who gambled, and Baderoon was guilty of both faults.

Nevertheless, Baderoon had moments of success:

> I was beginning to despair [of finding birds-of-paradise] [when] my boy Baderoon returned one day with a specimen which repaid me for months of delay and expectation.[30]

That exceptional bird was the king bird-of-paradise, one of the most impressive of the glorious birds-of-paradise, a trophy about which Wallace enthused:

> Thus one of my objects in coming to the far East was accomplished.[31]

But Baderoon, in spite of his collecting success, was never going to meet Wallace's expectations; he quit in distant Aru:

> My Macassar [now Makassar, Sulawesi] boy, Baderoon, took his wages and left me, because I scolded him for laziness. He then occupied himself in gambling, and at first had some luck, and bought ornaments, and had plenty of money. Then his luck turned; he lost every thing, borrowed money and lost that, and was obliged to become the slave of his creditor till he had worked out the debt. He was a quick and active lad when he pleased, but was apt to be idle, and had such an incorrigible propensity for gambling that it will very likely lead to his becoming a slave for life.[32]

And then there was Ali. This is his story, which is entwined with Wallace's own exceptional adventure. The simple takeaway moral: Wallace taught Ali. Ali taught Wallace.

Ali was a pretty good manager of camp logistics. Was he also a good manager of people? Did he take, or was he given responsibility for hiring local laborers and collectors? Did he give Wallace his opinions about whether a prospective laborer was energetic or lazy? Did he gossip with the other workers about his boss? We might imagine Ali sulking when Wallace hired yet one more collecting assistant. Particularly if that assistant was friendly with the boss.

He pays more attention to this new guy than to me, Ali thought. This kid doesn't know anything. He couldn't even chop open a durian if his life depended on it. He hasn't proved his loyalty. He gossips. Surely Tuan Wallace can see that?

And so the sabotage began.

CHAPTER 4

Ali and Wallace – In Sickness and in Health

Much has been made of Ali's nursing skills. Yet Wallace wrote that he, in turn, cared for Ali on various occasions when the young man was incapacitated by often-serious illnesses and injuries.

They helped each other overcome continuing episodes of fevers, inflammations, suppurating sores, relentless ticks, debilitating headaches, accidents, shipwrecks, and general misery, not to mention the exasperation of trying to manage a string of often-unreliable, sometimes-larcenous, hired hands.

Wallace Nursing Ali

While on the island of Lombok, for example, Wallace asked his host for "a horse for Ali, who was lame." Apparently the horse never appeared, and Wallace gave the young man his own mount:

> I gave Ali my horse, and started on foot, but he afterward mounted behind Mr. Ross's groom, and we got home very well, though rather hot and tired.[33]

In another passage, from Macassar, Wallace wrote about his concern for Ali's health, mixed with his annoyance at not having a regular cook:

Photo: Natalie Ong. The statues, located at the Lee Kong Chian Natural History Museum, Singapore, were sculpted by Chang Ting Hsuan, Lim XingYi, Subashri Sankarasubramanian, and Lim Soo Ngee of the Nanyang Academy of Fine Arts.

They were not equals, but Ali helped and cared for Wallace and vice versa. It is likely that Wallace would not have been as successful as he was without Ali's support.

Although this was the height of the dry season, and there was a fine wind all day, it was by no means a healthy time of year. My boy Ali had hardly been a day onshore when he was attacked by fever, which put me to great inconvenience, as at the house where I was staying nothing could be obtained but at meal-times. After having cured Ali, and with much difficulty got another servant to cook for me, I was no sooner settled at my country abode than the latter was attacked with the same disease, and, having a wife in the town, left me. Hardly was he gone than I fell ill myself, with strong

> intermittent fever every other day. In about a week I got over it by a liberal use of quinine, when scarcely was I on my legs than Ali again became worse than ever. His fever attacked him daily, but early in the morning he was pretty well, and then managed to cook me enough for the day. In a week I cured him.[34]

And Wallace sought medical help for Ali from a German doctor in Maros, north of Macassar (and again bemoaned the inconvenience):

> My boy Ali was so ill with fever that I was obliged to leave him in the hospital, under the care of my friend the German doctor, and I had to make shift with two new servants utterly ignorant of everything.[35]

And again in Macassar, Ali once more came down with fever, and again Wallace was annoyed that his routine was disrupted:

> My Malay boy Ali was affected with the same illness, and as he was my chief bird-skinner I got on but slowly with my collections.[36]

Ali Nursing Wallace

Wallace was prone to accidents and ill health and often suffered sores, bruises, and fevers.

Like other Victorian explorers, adventurers, mountain climbers, colonialists, missionaries, and fortune-seekers, Wallace didn't hesitate to tell his readers, always in an understated "aw shucks, it's really nothing" manner, how greatly he had suffered in the pursuit of his quest.

We might guess that his injuries were partly due to his stature – he was a six-foot-one-inch man living in an environment where humans were considerably shorter and more agile. He subjected himself to rough living conditions. And he had little natural immunity against pervasive jungle fevers, from which we suffered while in Brazil.

Wallace wrote about his aches and pains numerous times, and we can imply that Ali nursed him during these periods:

> All the time I had been in Ceram I had suffered much from the irritating bites of an invisible acarus [mite or tick], which is worse than mosquitoes, ants, and every other pest, because it is impossible to guard against them. This last journey in the forest left me covered from head to foot with inflamed lumps, which, after my return to Amboyna [Ambon], produced a serious disease, confining me to the house for nearly two months.[37]

Wallace spent a frustrating three and a half months, from April to July 1858, near Dorey (now Manokwari) in what is now the Indonesian province of West Papua (formerly Irian Jaya) on the island of New Guinea. In Dorey, Wallace found few insects, but he injured his ankle, which turned septic:

> [The wound] had to be leeched, and lanced, and doctored with ointments and poultices for several weeks.[38]

As a result he was frustrated and unable to move without a crutch:

> I was almost driven to despair — for the weather was at length fine, and I was tantalized by seeing grand butterflies flying past my door, and thinking of the twenty or thirty new species of insects that I ought to be getting every day. [39]

He suffered in Waigiou:

> Having been already eight months on this voyage, my stock of all condiments, spices and butter, was exhausted, and I found it impossible to eat sufficient of my tasteless and unpalatable food to support health. I got very thin and weak, and had a curious disease known (I have since heard) as brow-ague [Dr Grégory Fleury, a

rheumatologist at Hôpital de La Tour in Geneva, suggests Wallace might have been suffering from a cluster headache or, possibly, Horton's disease, also known as temporal arteritis]. Directly after breakfast every morning an intense pain set in on a small spot on the right temple. It was a severe burning ache, as bad as the worst toothache, and lasted about two hours... When this finally ceased, I had an attack of fever, which left me so weak and so unable to eat our regular food, that I feel sure my life was saved by couple of tins of soup which I had long reserved for some such extremity.[40]

Wallace was too incapacitated to trek into the birds-of-paradise-rich Arfak Mountains outside Dorey, and noted ruefully what he was missing in the steep and rugged hills:

> A few miles in the interior, away from the recently elevated coralline rocks and the influence of the sea air, a much more abundant harvest might be obtained.[41]

Poor Wallace. Sometimes he reminds me of the Peanuts character Pig-Pen, always enveloped in a dust cloud (or in his case, a non-stop infestation of ants):

> We bade adieu to Dorey, without much regret, for in no place which I have visited have I encountered more privations and annoyances. Continual rain, continual sickness, little wholesome food, with a plague of ants and flies, surpassing any thing I had before met with, required all a naturalist's ardor to encounter; and when they were uncompensated by great success in collecting, became all the more insupportable. This long-thought-of and much-desired voyage to New Guinea had realized none of my expectations. [But] Dorey was very rich in ants. They immediately took possession of my house... they swarmed on my table, carrying [my insect collection] off from under my very nose... they crawled continually over my hands and face, got into my hair, and roamed at will over my whole body...

> they visited my bed also... and I verily believe that during my three and a half months' residence at Dorey I was never for a single hour entirely free from them.[42]

And on the same trip, his bird skins were ravaged by flies:

> The flies that troubled me most were a large kind of blue-bottle or blow-fly. These settled in swarms on my bird skins when first put out to dry, filling their plumage with masses of eggs, which, if neglected, the next day produced maggots. They would get under the wings or under the body where it rested on the drying-board, sometimes actually raising it up half an inch by the mass of eggs deposited in a few hours; and every egg was so firmly glued to the fibres of the feathers, as to make it a work of much time and patience to get them off without injuring the bird. In no other locality have I ever been troubled with such a plague as this.[43]

The ants and flies were an annoyance, but it was a bout of malaria that changed Wallace's life.[44]

Wallace was most likely on the island of Halmahera, not far from Ternate, in February 1858, when he had arguably the most eventful malaria attack ever recorded:

> I was suffering from a sharp attack of intermittent fever, and every day during the cold and succeeding hot fits had to lie down for several hours, during which time I had nothing to do but to think over any subjects then particularly interesting me. One day something brought to my recollection Malthus's "Principles of Population" ["An Essay on the Principle of Population"], which I had read twelve years before. I thought of his clear exposition of "the positive checks to increase" – disease, accidents, war, and famine – which keep down the population of savage races to so much lower an average than that of more civilized peoples. It then occurred to me that these causes or their equivalents are continually acting in the case of animals

> also; and as animals usually breed much more rapidly than does mankind, the destruction every year from these causes must be enormous in order to keep down the numbers of each species, since they evidently do not increase regularly from year to year, as otherwise the world would long ago have been densely crowded with those that breed most quickly. Vaguely thinking over the enormous and constant destruction which this implied, it occurred to me to ask the question, Why do some die and some live? And the answer was clearly, that on the whole the best fitted live. From the effects of disease the most healthy escape; from enemies, the strongest, the swiftest, or the most cunning.... Then it suddenly flashed upon me that this self-acting process would necessarily *improve the race*, because in every generation the inferior would inevitably be killed off and the superior would remain-that is, *the fittest would survive* ... I waited anxiously for the termination of my fit so that I might at once make notes for a paper on the subject [italics Wallace]. [45]

This, as they say, was eureka time. Archimedes in his bath, Newton under the apple tree. Wallace in a cold sweat.

Wallace recognized the importance of his idea. Once he got back his strength and returned to his base camp in Ternate, he wrote, over two evenings, a paper he titled "On the Tendency of Varieties to Depart Indefinitely from the Original Type," in which he outlined the mechanism of natural selection; Wallace later described his work as "the long-sought-for law of nature that solved the problem of the origin of species."[46]

Wallace sent the paper to Charles Darwin and asked him to show it to noted geologist Charles Lyell. Up to that point Darwin had not published a single word on evolution. While most modern chroniclers recognize Darwin and Wallace as co-discoverers of the Theory of Evolution by Natural Selection, several commentators, including David J. Hallmark, a Worcester, UK-based lawyer, argue that Darwin plagiarized Wallace's theory,[47] but that's a conversation for other books, other debates.[48] [49]

This episode of malaria-induced genius is widely recognized.

The BBC made a cringe-generating film about Alfred Russel Wallace. Here's how they handled the "malaria fit" apotheosis:

> Wallace is lying feverish in bed. An ethereal woman's voice off camera whispers: "Where do they all come from Alfred? Out of the Ark?"
>
> Wallace, weak, hesitant: "Always where there is a closely related, pre-existing species."
>
> Aggressive male voice: "You still haven't explained how one species can change into another."
>
> Wallace, a bit stronger: "The individual deaths don't matter. It's as Malthus said, but nobody knows the mechanism."
>
> Darwin's voice: "I agree with everything you wrote about the genesis of species, but I would go further, much further."
>
> Wallace, stronger yet: "It doesn't matter as long as some of them survive, the strongest ones or the ones that can run fastest, or the ones best fitted to the kind of..." At this point Wallace opens his eyes, and there is a trumpet fanfare [true]. "That's it!!" Wallace calls to his faithful servant Ali: "Pencil and paper!"
>
> Ali sees that his master is still delirious. "No, Tuan. Tomorrow."

Wallace's malaria dream was generally known by Wallace's contemporaries and supporters, and American naturalist Thomas Barbour refers to it following his meeting Ali in Ternate (see Chapter 13, "Ali Returned to Ternate?"):

> [Wallace] wrote me a delightful letter acknowledging it and reminiscing over the time when Ali had saved his life, nursing him through a terrific attack of malaria.[50]

CHAPTER 5

Ali's Career Path – Camp Manager

Wallace described how Ali grew on the job. He started as a cook, learned to collect birds (generally by shooting them), and quickly picked up the techniques of preparing bird skins.

Ali was a stable, reliant assistant:

> In all the difficulties or dangers of our journeys [Ali] was quite undisturbed and ready to do anything required of him.[51]

But the most dramatic career advance was that Ali, still a teenager, grew sufficiently in stature to instruct, and later to help manage, the assorted cooks, camp assistants, boat builders, and laborers whom Wallace relied on:

> He was a good boatman, as are all Malays . . . He accompanied me through all my travels, sometimes alone, but more frequently with several others, and was then very useful in teaching them their duties, as he soon became well acquainted with my wants and habits.[52]

Travelling and moving can be stressful. Even today, packing for a short vacation where you will stay in a comfortable hotel might lead to anxiety about what to pack. Consider then, the exhausting challenges Wallace faced when he wanted to move camp.

For starters, many of Wallace's camps were located on relatively inaccessible islands. Get a good atlas and see if you can find the route between Manawoka and Watubela islands. You can't pop down to the local hardware store if you forget to bring the nails or saw.

And all the goods often had to be carried in small boats. Unfortunately for Wallace, he was terrified of sailing in such craft. He had awful memories of when the ship he was travelling in, the brig *Helen*, caught fire in the Atlantic, some 700 miles east of Bermuda. He watched in despair as much of his Amazon collection and the majority of his notes burned and sank into the ocean.

> And now everything was gone, and I had not one specimen to illustrate the unknown lands I had trod.[53]

He suffered for ten days aboard a rowboat hoping to be rescued. With typical understatement he wrote:

> Day after day we continued in the boats. The winds changed, blowing dead from the point to which we wanted to go. We were scorched by the sun, my hands, nose, and ears being completely skinned, and were drenched continually by the seas or spray. We were therefore almost constantly wet, and had no comfort and little sleep at night. Our meals consisted of raw pork and biscuit, with a little preserved meat or carrots once a day, which was a great luxury, and a short allowance of water, which left us as thirsty as before directly after we had drunk it.[54]

On his return to England he swore never to go to sea again:

> Fifty times since I left Para [now Belém] have I vowed, if I once reached England, never to trust myself more on the ocean.[55] [56]

If he was lucky he could book passage on a government sailing vessel or a trader's ship.

But often he was left to his own devices.

He hired boats but had to find his own crew; he describes one such excursion trying to go from Ternate to Batchian (Bacan):

> I found that I should have to hire a boat ... I accordingly went into the native town, and could only find two boats for hire, one much larger than I required, and the other far smaller than I wished. I chose the smaller one, chiefly because it would not [*sic*] cost me one-third as much as the larger one, and also because in a coast-voyage a small vessel can be more easily managed, and more readily got into a place of safety during violent gales, than a large one.[57]

Sometimes he was forced to build (and find crew for) his own boat:

> My experiences of travel in a native prau have not been encouraging. My first crew ran away; two men were lost for a month on a desert island; we were ten times aground on coral reefs; we lost four anchors; the sails were devoured by rats; the small boat was lost astern; we were thirty-eight days on the voyage home [to Ternate], which should not have taken twelve; we were many times short of food and water; we had no compass-lamp ... we had *not one single day of fair wind!* ... Every seaman will admit that my first voyage in my own boat was a most unlucky one [italics Wallace].[58]

Wallace does not mention Ali specifically, but one can imagine how Ali assisted in explaining Wallace's desires to the local bird hunters:

> My first business [on arrival at Waigou] was to send for the men who

> were accustomed to catch the birds of paradise. Several came … and [I] explained to them, as well as I could by signs, the price I would give for fresh-killed specimens. After three days, my man brought me the first bird – a very fine specimen but tied up in a small bag, and consequently its tail and wing feathers very much crushed and injured. I tried to explain to him … that I wanted them as perfect as possible … As they caught them a long way off in the forest, they would scarcely ever come with one, but would tie it by the leg to a stick, and put it in their house till they caught another. The poor creature would make violent efforts to escape, would … hang suspended by the leg till the limb was swollen and half putrefied, and sometimes die of starvation and worry … Luckily, however, the skin and plumage of these birds is so firm and strong, that they bear washing and cleaning better than almost any other sort.[59]

And Ali's support might also be inferred when Wallace complains about the hardships he endured while searching for specimens:

> I determined, therefore, to stay as long as possible, as my only chance of getting a good series of specimens; and although I was nearly starved, every thing eatable by civilized man being scarce or altogether absent, I finally succeeded.[60]

These hardships were made even more miserable by his frequent hunger, which he sated stoically. In extremis, he ate the flesh of his birds-of-paradise after skinning them because he had no other provisions, but complained about the culinary experience:

> The flesh … is dry, tasteless, and very tough – to be eaten only in necessity.[61]

Wallace collected at some 100 locations and required the services of up to 13 men to move and establish base camps. Having a camp manager like Ali to organize these voyages was invaluable.

Yochahile Haruna, a skilled boat-builder of the Aru islands, working with traditional craft techniques.
Paul Spencer Sochaczewski.

There were some situations even Ali couldn't fix. Wallace wrote in *The Malay Archipelago* about his frustrations with boat-builders he had hired: "Five men had been engaged to work. Their ideas of work were, however, very different from mine, and I had immense difficulty with them; seldom more than two or three coming together, and a hundred excuses being given for working only half a day when they did come. Yet they were constantly begging advances of money, saying they had nothing to eat. When I gave it [to] them they were sure to stay away the next day, and when I refused any further advances some of them declined working any more."

John Bastin, co-author of *A History of Modern Southeast Asia*, writes: "[Wallace] had to make his own travelling arrangements, lay in provisions and supplies for himself and his assistants, and arrange for the transportation of his tents, camp-beds folding chairs and tables, cooking utensils, collecting nets, bottles, chemicals, insect boxes, display boards, guns, ammunition, prismatic compasses, thermometers, barometers, a sympiesometer as well as other baggage."[62]

To this packing list we might add Wallace's various notebooks, lanterns and construction tools; "a cask of medicated arrack to put mias [orangutan] skins in"; dissecting and taxidermy equipment; food; medicines; "a great iron pan" for boiling orangutan carcasses to get a commercially viable skeleton; a locked box with coins for buying supplies and paying helpers; "hatchets, beads, knives, and handkerchiefs" to be used as payment for "fresh-killed specimens." He also carried several heavy books: Charles-Lucien Bonaparte's *Conspectus Generum Avium* (Leiden, 1850), Jean Alphonse Boisduval's *Histoire Naturelle des Insects* (Paris, 1836), and Charles Lyell's *Principles of Geology* (first published in London, 1830), the only book carried by both Wallace (in Southeast Asia) and Darwin (during his voyage of the Beagle).

How much easier Wallace's life would have been if he had a dozen sets of 21st-century reusable air-tight kitchen containers (Tupperware!), several rolls of duct tape, large plastic storage boxes, a couple of large heavy-duty plastic sheets, spray insecticide for specimens, mosquito repellent, a modern medical kit with antibiotics, anti-fungal treatments, malaria prophylactics, sterile bandages, some ballpoint pens, and an inflatable mattress.

Wallace trusted Ali with large amounts of cash. As an aside in a long description of a robbery, Wallace wrote that he sent Ali to Ternate to collect a bag of coins:

> Soon after I arrived here [the island of Batchian, now Bacan], the Dutch Government introduced a new copper coinage of cents instead of doits (the 100th instead of the 120th part of a guilder), and all the old coins were ordered to be sent to Ternate to be changed. I sent a bag containing 6,000 doits, and duly received the new money by return of the boat. *When Ali went to bring it*, however, the captain required a written order; so I waited to send again the next day, and it was lucky I did so, for that night my house was entered, all my boxes carried out and ransacked, and the various articles left on the road about twenty yards off, where we found them at five in the morning, when, on getting up and finding the house empty, we rushed out to discover tracks of the thieves [italics added].[63]

Within a year of hiring him Wallace described Ali as "my head man":

> Ali, the Malay boy whom I had picked up in Borneo, was my head man. He had already been with me a year, could turn his hand to any thing, and was quite attentive and trustworthy. He was a good shot, and fond of shooting, and I had taught him to skin birds very well.[64]

And Wallace eventually trusted Ali to go by himself to buy natural history specimens. Note that Ali was ill (poor health dogged both Ali and Wallace throughout the journey) and that he was trustworthy and respected by outsiders:

> My boy Ali returned from Wanumbai, where I had sent him alone for a fortnight to buy Paradise birds and prepare the skins; he brought me sixteen glorious specimens, and had he not been very ill with fever and ague might have obtained twice the number. He had lived with the people whose house I had occupied, and it is a proof of their goodness, if fairly treated, that although he took with him a quantity of silver dollars to pay for the birds they caught, no attempt was made to rob him, which might have been done with the most perfect impunity. He was kindly treated when ill, and was brought back to me with the balance of the dollars he had not spent.[65]

And in due course Wallace trusted Ali to scout locations that might be sufficiently productive to justify going to the expense and trouble of moving his camp:

> It became evident, therefore, that I must leave Cajeli [on Buru] for some better collecting-ground ... *I sent my boy Ali ... to explore and report* on the capabilities of the district [Pelah] ... [after an account of a difficult walk that occupies two pages of text, Wallace continues] I waited Ali's return to decide on my future movements. He came the following day, and gave a very bad account of Pelah, where he had been [italics added].[66]

Imagined Conversation: "He Raises the Dead!"

Ali and Wallace are camped on a small island in eastern Indonesia that is seldom visited by white men. Earlier in the day Wallace had shown some local men what a butterfly he had collected looked like under a magnifying glass. That experience became the subject of conversation for hours.

After preparing dinner for Wallace (who often eats alone), Ali goes to a village house for a simple meal of rice, boiled vegetables, and a weak soup of river fish rife with small bones and spiced only with fiery chilies. He has liberated some roast chicken, which he had prepared for Wallace, and shares it with the men.

"Your man," one of the villagers says. "He's a shaman?"

Ali asks what he means.

"He collects these creatures and uses them for spells and incantations."

"How do you know that?"

"Well, it's obvious. Why else would anyone bother to collect all these things unless he was going to use them to heal people?"

"Or put a spell on them!" another man adds.

"Or raise the dead!" a third man insists. "I heard he brings the dead back to life. I spoke with Tuan Wallace yesterday, when I brought him some beetles. He laughed when I told him that I knew his secret, but he just ignored me. But that's true, isn't it, Young Brother Ali? He makes the dead come alive!"

"You're just simple village men," Ali says. "If you've seen what I've seen, you wouldn't make up stories like that. Have you ever heard of a place called Singapore?"

"Tuan Wallace stays up all night; I've seen his lamp shining. That's when he does his magic!" another villager says with authority.

Yet another villager adds: "Does he use magic to grow his beard? Can you get me some of the potion? My wife would like me to have a big beard like that. Very strong!" Laughter.

"Stop this stupid talk. Bring me some beautiful birds and butterflies tomorrow. I'll pay you well," Ali says, keen to change the subject.

"And he says he's from a place called 'An-glung,'" the first villager says, refusing to drop the subject. "Where is that, Ali?

"Far across the ocean."

"How many days paddling?"

"In his country Tuan Wallace is a great pawang," Ali says, his voice taking on an authority that doesn't match his short stature. "In his country there are bears the size of this house and crocodiles as long as your prahu. Whenever there's a problem, the officials send for him, and he sings a pantun into the animal's ear to get the beast to repent and follow the Christian god."

Much consternation and whispering.

"So, my advice is, when Tuan Wallace asks you to get specimens, don't try to fool him. And never, ever cheat or steal from him. He knows exactly what you are thinking."

More concerned murmuring.

"And yes, he talks to the dead."

Yet more disquiet and alarm, accompanied by copious consumption of rice wine.

"You know, in Singapore I shot a tiger," Ali says, looking each man in the eyes. "A huge man-eating tiger. The governor invited me to his palace; he's the top white man there, even more important than Tuan Wallace. His astana was as big as this entire island. He had a zoo with giant horrible creatures they say came from outer space. All the furniture was made of gold. He has an army of ten thousand men, and the same number of vicious women bodyguards who will cut off your balls if you insult him. He gave me a medal..."

CHAPTER 6

Ali's Career Path – Bird Hunter, Taxidermist

Ali and other collectors were key to Wallace's success as a bird hunter.

The statistics are impressive: Wallace is credited with collecting 8,050 bird specimens. After subtracting the estimated 2,900 birds collected by Charles Allen, the young man who was Wallace's first senior assistant, John van Wyhe and Jerry Drawhorn estimate that "Ali probably collected many of the remaining 5,150 birds,"[67] an astonishing total. A remarkable 212 of the species were new. Wallace named 105 of the new species while 23 of the new species and subspecies he collected were named after him by other taxonomists, including the new bird-of-paradise shot by Ali – *Semioptera wallacii*.[68]

Wallace reveled in the beauty and unusual behavior of many of the wild birds he subsequently shot, stuffed, and shipped off to England.

In Dodinga, Gilolo (Halmahera), he wrote:

> I got some very nice insects here, though, owing to illness most of the time, my collection was a small one; and my boy Ali shot me a pair of one of the most beautiful birds of the East (Pitta gigas), a large ground-thrush, whose plumage of velvety black above is relieved by a breast of pure white, shoulders of azure

> blue, and belly of vivid crimson. It has very long and stong [*sic*] legs, and hops about with such activity in the dense tangled forest, bristling with rocks, as to make it very difficult to shoot.[69]

Ali showed that he had pride in his work. It was more than a job. He seemed to enjoy his role as a trusted bird collector (a cousin of Papageno?) and was willing to endure pain and discomfort to obtain a rare creature:

> Soon after we had arrived at Waypoti, Ali had seen a beautiful little bird of the genus Pitta, which I was very anxious to obtain, as in almost every island the species are different, and none were yet known from Bouru [now Buru]. He and my other hunter continued to see it two or three times a week, and to hear its peculiar note much oftener, but could never get a specimen, owing to its always frequenting the most dense thorny thickets, where only hasty glimpses of it could be obtained, and at so short a distance that it would be difficult to avoid blowing the bird to pieces. *Ali was very much annoyed that he could not get a specimen of this bird*, in going after which he had already severely wounded his feet with thorns; and when we had only two days more to stay, *he went of his own accord* one evening to sleep at a little hut in the forest some miles off, in order to have a last try for it at daybreak, when many birds come out to feed, and are very intent on their morning meal. The next evening he brought me home two specimens, one with the head blown completely off, and otherwise too much injured to preserve, the other in very good order, and which I at once saw to be a new species, very like the Pitta celebensis, but ornamented with a square patch of bright red on the nape of the neck [italics added].[70]

Ali was not only a skilled bird hunter, but he learned taxidermy, a highly useful skill. Wallace made his living selling "natural productions" to collectors and museums in Europe. The bird specimens, if they were to have commercial value, had to be skinned, cleaned, dried, and labeled, and adequately packed to survive a multi-month shipment to London. Wallace wrote simply:

> I had taught him to skin birds very well [and Ali] soon learnt to shoot birds, to skin them properly, and latterly even to put up the skins very neatly.[71]

This specialist skill may well have been a factor in Ali's desire to return to Ternate after Wallace left for England because Ali could likely have found employment as a taxidermist with Maarten Dirk van Renesse van Duivenbode (Wallace misspelled his name as van Duivenboden), whom Wallace said was called the "king of Ternate" due to his wealth and influence, writing: "He was a very rich man, owned half the town, possessed many ships, and above a hundred slaves." Van Duivenbode, who helped Wallace get settled in Ternate, was himself a prominent dealer in exotic birds, particularly birds-of-paradise (see Chapter 13, "Ali Returned to Ternate?").

Imagined Conversation: "I Want to Go with You"

Wallace and Ali are going over the day's take — a few dozen beetles (some as tiny as a grain of rice, a few large and garish, and one giant with frightening pincers), a handful of glorious butterflies, and a gray parrot.

"Ali, show me how you skin this parrot," Wallace says in Malay.

Ali tries not to giggle. He always giggles when Wallace speaks Malay. It sounds like cats fighting. "By myself?"

"You're the only Ali here."

So, Ali sharpens the scalpel, as he has seen Wallace do countless times.

"Separate the feathers on the ventral part of the body — that's the bottom part of the bird. Start to cut at the throat and slice down to the cloaca — all the way at the bottom."

Ali feels proud that he knows what Wallace means when he uses scientific words.

"Slowly, Ali. Better to make one clean cut than several sloppy cuts. Slowly-slowly catch the monkey."

"You know that saying, Tuan?" Ali asks.

"It's English."

"No, it's Malay." Ali slices the bird's breast and pins the carcass on the workbench — a grand name for a couple of planks — to expose the bird's organs.

"Careful. Put some kapok into its body to absorb the blood. You don't want to ruin the color of the feathers."

Ali continues as Wallace hovers. The young man removes the bird's organs and is about to throw them through a hole in the rattan floor. Below, scavenger dogs smell a potential snack and jostle for position below the shack on stilts that is Wallace's home, laboratory, study, and store room.

"Be careful. Don't throw the guts on the floor."

Ali is confused. "Where should I put it, then?"

Wallace is irritated. He brusquely takes the scalpel, lays the guts on a clean space, and slices the crop open. "Look here, Ali, what do we see?"

Wallace doesn't wait for an answer. "We can determine what the bird was eating when you so cruelly shot it."

Ali comes from a culture where irony is still in its infancy; he doesn't reply.

Wallace picks out half-digested critters. "This looks like a caterpillar. This was a beetle, probably. And what's this gloop?" he asks, poking in a mess of jelly-like stomach contents. He rubs his fingers in the soft mixture and smells it, then tastes it, to Ali's disgust.

Wallace wipes his hands and writes in his notebook, mumbling to himself as he does so: "Caterpillar. Beetle. Bit of something, maybe a spider. And a half-digested bit of ripe fig, which would have been perfect for making confiture if this creature hadn't eaten it first. Ah, well, a bird's also got to survive."

Ali says nothing; he is used to the idiosyncrasies of the man he works for.

"And don't get blood on the workbench."

"It's just blood, Tuan."

"To you it's just blood. To the rats and the ants, it's a fiesta. And I don't care how *you* choose to sleep; I don't want to sleep with ants and rats crawling around."

"Why do you do this, Tuan?"

"You mean collecting?"

"All these useless insects and birds."

"Because people in Europe buy them. And that pays for your salary."

Europe. Ali has heard of Europe from conversations he had overheard when Wallace was speaking with other white men.

"Tell me about this 'Europe.'"

And Wallace tells Ali some stories about great cities, large buildings, women in fine dresses, the wonderful thing called the steam engine that was just becoming fashionable, and the miracle of photography."

"I would like a photograph, Tuan."

"Hmmm. Maybe when we get to Singapore, Ali. I think they have a photographer there."

Ali is silent. He continues working until the final step when he sprinkles arsenic into the empty carcass and adds some clean cotton. He speaks softly. "I want to go with you to England, Tuan."

It is Wallace's turn to remain silent. Finally, he says: "You wouldn't be happy there. You're better here, with your people."

"You're my people, Tuan. You taught me to cut the birds." And in English he says, "You teach me An-glush."

"Finish up this bird, Ali. Then I'm going to sleep. I'm very tired, and we have a long day tomorrow."[72]

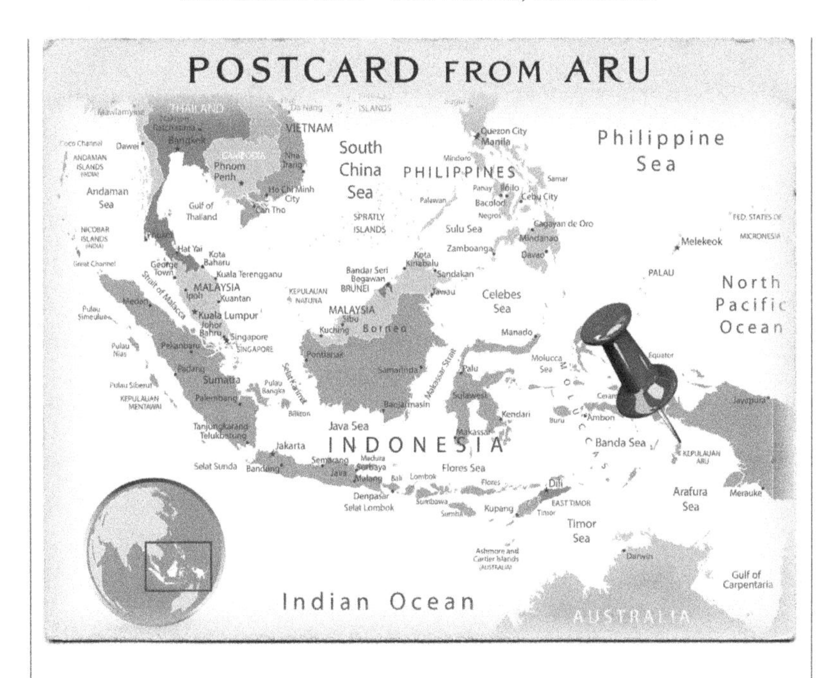

If you ask people from Western Indonesia if they have heard of Kayoa, Manawoka, or Watubela and if they can point them out on a map, chances are they will shrug.

But for Wallace, such obscure islands were sources of great adventure, insight, and profit.

Indonesia has some 17,000 islands, spanning roughly the distance of San Francisco to New York, and while it's easy to tick off your list the big, famous islands like Sumatra, Borneo, Sulawesi, Bali, or Flores, it takes effort, time, and sometimes money to get to places Wallace visited like Gorong, Waru, or Waigeo. Of course, life on these outposts has changed dramatically since Wallace's time. Most places now have some electricity and likely a primary school and a simple dispensary (availability of medicines or a trained nurse is not certain) and perhaps even a tiny airport. Some also have cellphone coverage and better inter-island transport. But when you visit many of these small islands, if you half-close your eyes and use your imagination, it's not hard to imagine a simpler, sustenance existence similar to what Wallace would have encountered.

One of those often-ignored lands that Wallace visited is the Aru archipelago, a group of some 95 islands roughly 50 miles by 110 miles in size where people speak 14 indigenous languages. I travelled there in 1990 with a Dutch conservation team.

At the time, Dobbo (now Dobo), the capital of Aru regency, was a dirty, hectic, unappetizing trading center. Dutch biologist Mark van der Wal and I were put up in a simple wooden guesthouse on stilts, rife with mosquitoes, noise, and odors. The flimsy structure was perched over the seafront that was littered with rubbish and the excretions of residents who, like us, used the overhanging toilet that was a simple hole in the wooden floor. I've never encountered a town that had such a mercantile focus on selling creatures, dead and alive, that came from the seas and forests.

In fairness, my visit to Aru was more than 30 years ago. A quick internet search shows that today a handful of tourists (mostly bird watchers who arrive on a weekly flight from Ambon) can stay in a comfortable air-conditioned hotel in town or in an eco-resort in the forest that offers "outstanding luxury."

Wallace wrote of the Aru archipelago:

> These islands [Aru] are quite out of the track of all European trade, and are inhabited only by black mop-headed savages, who yet contribute to the luxurious tastes of most civilized races. Pearls, mother-of-pearl, and tortoise-shell, and birds of paradise find their way to Europe, while edible birds' nests and 'tripang' [trepang], or sea-slug, are obtained by shiploads for the gastronomic enjoyment of the Chinese.[73]

Wallace was impressed by the scale of trade that occurred:

> The trade carried on at Dobbo is very considerable. This year [1857] there were fifteen large praus from Macassar, and perhaps a hundred small boats from Ceram, Goram, and Ké [Kei]. The Macassar cargoes are worth about £1000 each, and the other

boats take away perhaps about £3000 worth, so that the whole exports may be estimated at £18,000 per annum.[74]

Alfred Russel Wallace, The Malay Archipelago. *Dobbo in the trading season.*

One of the main trading centers of eastern Indonesia for interesting, commercial-quality birds was Dobbo (now Dobo). The infrastructure in the town has improved since Wallace's time, but the goods on sale — turtles, shark fins, mother-of-pearl, trepang, birds' nests, birds-of-paradise, parrots — haven't changed much over the years.

Mark and I chartered a small outboard in Dobo and travelled south to the Manoembai River.

A farmer and hunter named Ely invited us to his nearby home close to the tiny village of Jirlai on the island of Kobroor. We stayed in an unfinished house he was constructing in the forest. It was set over slow-moving fresh water, accessible by walking a plank and crawling through a window frame. It was empty, yet it was one of the most comfortable shelters I had had on that trip, particularly after two weeks of trying to sleep on the decks of small boats ill-fitted for that purpose. I glanced around and did a quick inventory: The only externally produced things Ely owned were cooking utensils, plastic buckets, his machete, a few gardening tools, a battery-powered cassette player, and a kerosene pressure lamp, plus a few clothes and one or two pictures of Indonesian movie stars ripped out of a magazine and nailed to the wall.

I asked Ely and his brother Yos how they made money. They rely on nature, and two of their main sources of cash — the birds-of-paradise and the swiftlets that make edible birds' nests — are protected or vulnerable species. In both cases, they earn their modest income by selling the same commercial wildlife as their ancestors did during Wallace's time in the mid-19th century.[75]

Alfred Russel Wallace, The Malay Archipelago

Natives of Aru shooting the great bird-of-paradise in the mid-19th century. The technique hasn't changed much since Wallace's time — hunters shoot a wooden-capped arrow to stun the bird, but without damaging the valuable plumage.

It's a tough life. Ely and Yos don't require a lot of money, but they need enough to buy supplies, pay the school fees for their children, travel to Dobo if someone gets sick, and buy a new sarong for the wife once in a while. They'd love to have an outboard engine and electricity. How ironic: In one sense Mark and I envied people like Ely and Yos for their simplicity. They envy us for our possessions. I thought of theologian Thomas Berry's comment that the future belongs not to those who have the most but to those who need the least. I bet Ely and Yos wouldn't agree. They see the present — and in the present, the person with the most toys wins.

Wikimedia Commons

The Aru regency coat of arms features a bird-of-paradise.

Had they noticed a reduction in birds or fish or big mammals?

"Yes. There are fewer birds' nests to collect now," the two men told me.

"But why?"

"We collect the nests three or four times a year, so there are fewer swiftlets, of course."

"What if you only collect nests twice a year? What if you set up some kind of control system?"

"Yes! *Sasi*," they said, referring to a traditional control of harvesting natural resources. But they gave me looks that said it will never work. "The problem is, if we don't take them, someone else will."

"Who?"

"Outsiders," they said cryptically. I pushed them, perhaps too aggressively, to be more specific. "Javanese," they explained. And people from other islands, working on commission from mostly Chinese traders.

Ely and Yos are stuck in a system that hasn't changed much since Wallace was here. Same game, different players. It's a vicious cycle. Wherever local people aren't in control of their resources, nature gets hammered.[76]

CHAPTER 7

Wallace, Ali, and the Search for Birds-of-Paradise

Alfred Russel Wallace's classic book of travel, natural history, and ethnography, *The Malay Archipelago* is subtitled: "The Land of the Orang-Utan and the Bird of Paradise."

Wallace was partly successful in his search for these two forms of iconic creatures.

He famously shot, skinned, and sold some 17 orangutans in what is now Sarawak, a Malaysian state on the island of Borneo. While Wallace earned good money from these specimens, he also studied the animals, alive and dead, and developed important insights about orangutan ecology and behavior.[77]

While other Western collectors and traders had preceded him in search of birds-of-paradise (a flourishing bird trade existed in Ternate), Wallace was a collector *and* naturalist. He paid attention to the behavior, ecological niches, and geographical differences of the creatures he caught, pinned, skinned, boiled, and pickled.

He was less successful in his search for birds-of-paradise. While he is credited with collecting 8,050 birds, he was disappointed that he only obtained a handful of bird-of-paradise species (of the 18 then known to exist), many of which were collected by Ali and other assistants.

On the island of Batchian (Bacan), Ali famously collected a bird that became one of Wallace's most-prized specimens:

Just as I got home I overtook Ali returning from shooting with some birds hanging from his belt. He seemed much pleased, and said, "Look here, sir, what a curious bird!" holding out what at first completely puzzled me. I saw a bird with a mass of splendid green feathers on its breast, elongated into two glittering tufts; but what I could not understand was a pair of long white feathers, which stuck straight out from each shoulder. Ali assured me that the bird stuck them out this way itself when fluttering its wings, and that they had remained so without his touching them. I now saw that I had got a great prize, no less than a completely new form of the Bird of Paradise, differing most remarkably from every other known bird. This striking novelty has been named by Mr. G.R. Gray of the British Museum, Semioptera Wallacei [today it's known as *Semioptera wallacii*], or "Wallace's Standard-wing."[78] [79]

Joseph Wolf – A monograph of the Paradiseidae or birds of paradise by Elliot, Daniel Giraud (1873) Biodiversity Heritage Library.

Significantly, this "curious" bird was the only new bird-of-paradise he collected.

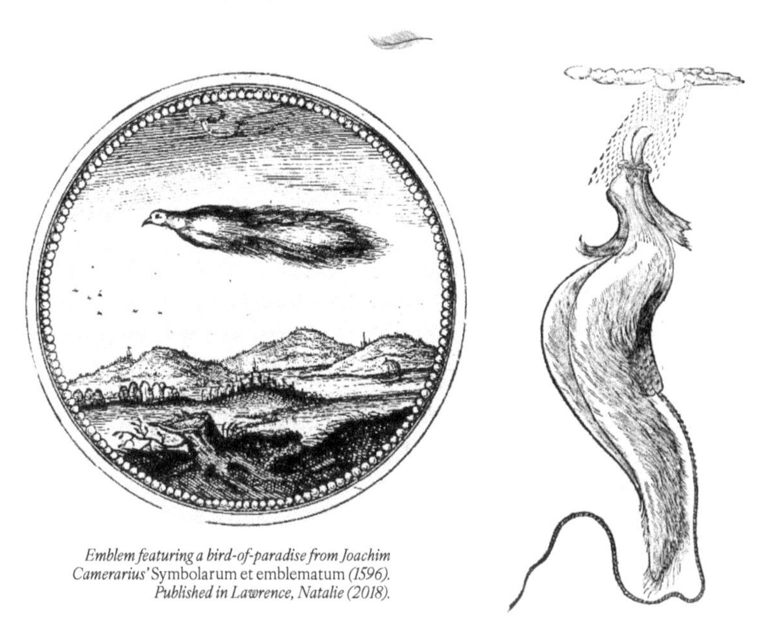

Emblem featuring a bird-of-paradise from Joachim Camerarius' Symbolarum et emblematum *(1596). Published in Lawrence, Natalie (2018).*

Bird-of-paradise shown here drinking rain, from Ulisse Aldrovandi's Ornithologiae *(1599). Published in Lawrence, Natalie (2018).*

How did the birds-of-paradise get their evocative name?

When Ferdinand Magellan's crew visited the island of Tidore in 1521, they were offered a gift of beautiful dead birds by the Sultan of Bacan to give to Charles I, the king of Spain. Based on the account of the voyage by Magellan's assistant Antonio Pigafetta, and considering the circumstances and description of the birds, they were likely standardwings, the type of bird-of-paradise collected by Ali and presented to Wallace.

The term itself might have originated when Maximilianus Transylvanus, who chronicled Magellan's voyages, introduced the name manucodiata (a corruption of the Malay manute-dewata, "bird of the gods").

The most mundane explanation of this evocative name is that early Portuguese and Dutch spice-seekers were presented with brightly colored bird skins that had no feet. The locals had simply hacked off the appendages, perhaps deciding that a legless stuffed bird was more aesthetically pleasing and, therefore, worth more in trade. The imaginative Europeans, however, deduced that the creatures were legless because they spent their entire lives in the air, survived by drinking rain, and used a depression in the female's back as an aerial nest. The Portuguese dubbed them Passaros de Sol, or Birds of the Sun. The Dutchmen who followed the Portuguese named the creature Avis Paradiseus, or Paradise Bird.

The belief that these birds had no feet was consolidated by Carl Linnaeus, the founder of the modern system of binominal classification, who described the greater bird-of-paradise in 1758 in his System Naturae ("The System of Nature"). He named it Paradisaea apoda, which translates as legless bird-of-paradise.

How did Wallace identify this bird as a new species? He had little formal education, having left school at the age of 14, and he did not come from a scientifically oriented family. His biological knowledge was largely self-taught.

He travelled with a single bird book, Charles-Lucien Bonaparte's *Conspectus Generum Avium*, published in 1850. George Beccaloni, a senior entomologist and science historian who created and manages several important Wallace-related websites, notes "rather incredibly it contains only brief Latin descriptions of the birds and lacks any illustrations. Wallace obviously had a good knowledge of Latin and a remarkable grasp of the distinguishing characteristics of birds, since using the brief descriptions in this book he was able to visualize exactly what the species looked like. He later claimed that thanks to this book he 'could almost always identify every bird already described, and if I could not do so, was pretty sure that it was a new or undescribed species.'" [80] [81]

Upon realizing he had a new treasure in his hands, a bird perhaps new to Western science, he excitedly wrote to his agent Samuel Stevens in London. Wallace was so excited he used an inordinate number of exclamation points, likely setting a personal record for enthusiastic punctuation:

> I believe I have already the finest & most wonderful bird in the island [Batchian, now Bacan, located south of Ternate]. I had a good mind to keep it a secret but I cannot resist telling you. I have got here a new *Bird of Paradise!!* of a new genus!!! quite unlike any thing yet known, very curious & very handsome!!! When I can get a couple of pairs I will send them overland to see what a new Bird of Paradise will really fetch, — I expect £25 each! [82] [83]

Besides being of great beauty (and significant potential income), Wallace also conveyed the news that this was an ornithologically important discovery, proving that birds-of-paradise were found far further north and west of the Maluku islands (also known as the Moluccas) than previously documented:

> Had I seen the bird in [a market in] Ternate [where birds-of-paradise

> from the southern region of Maluku were actively traded], I would never have believed it came from here, so far out of the hitherto supposed region of the Paradiseidae.[84]

And, without thanks to Ali, who goes unmentioned in this exchange, Wallace claimed this was his "discovery":

> I consider it the greatest discovery I have yet made & it gives me hopes of getting other species in *Gilolo & Ceram* [italics Wallace].[85]

And in *The Malay Archipelago* he also boasts about his find, again neglecting to mention Ali:

> The Standard Wing, named Semioptera wallacei by Mr. G. R. Gray, is an entirely new form of Bird of Paradise, *discovered by myself* in the island of Batchian [italics added].[86]

He wrote a detailed description of the bird, which he sent to noted ornithologist John Gould (who first identified the birds that later became known as "Darwin's finches") of the Zoological Society of London:

> The *Semioptera wallacii* frequents the lower trees of the virgin forests, and is almost constantly in motion. It flies from branch to branch, and clings to the twigs and even to the vertical smooth trunks almost as easily as a Woodpecker. It continually utters a harsh croaking cry, something between that of *Paradisea apoda* [greater bird of paradise] and the more musical cry of *Cicinnurus regius* [king bird-of-paradise]. The males, at short intervals, open and flutter their wings, erect the long shoulder feathers, and expand the elegant shields on each side of the breast. Like the other Birds of Paradise, the females and young males far outnumber the fully plumaged birds, which renders it probable that the extraordinary accessory plumes are not fully developed until the second or third

> year. The bird seems to feed principally upon fruit, but it probably takes insects occasionally.
>
> The iris is of a deep olive; the bill horny-olive; the feet orange, and the claws horny.[87]

Wallace sent a written description and what he described as "a terrible sketch" of the bird to his agent Samuel Stevens, who then forwarded this information to George Robert Gray of the British Museum. Working from the notes and sketch, but without having seen a specimen, Gray named it *Paradisea (Semioptera) wallacei* in Wallace's honor (it is now *Semioptera wallacii*). Frustratingly, Wallace's sketch has disappeared, one more mystery of important items missing from Wallace's archives. There was the loss of most of his South American collection when the brig *Helen* caught fire and sank; his lost letter to Darwin from Ternate with the attached Ternate Essay manuscript; the manuscript itself (sometimes also referred to as the Ternate Letter or the Ternate Paper), presumably lost at the printer; the loss of 25 of the 57 vocabularies he gathered from the Malay Archipelago ("distinct languages . . . more than half of which I believe are quite unknown to philologists," Wallace wrote), which he had lent to a friend who lost them while moving house ("they probably found their way to the dust-heap along with other waste paper," Wallace wrote); and the tantalizing mystery of the purported photo of Ali that American naturalist Thomas Barbour said he took and had sent to Wallace — about which Wallace said that Barbour had actually sent a photo of tribesmen from New Guinea, but he would have preferred to have received a photo of Ali. (For more about Barbour's encounter with Ali in Ternate, see Chapter 13, "Ali Returned to Ternate?")

In spite of his enthusiasm, Wallace was disappointed in his search for birds-of-paradise.

Partly this was due to his debilitating ankle injury while residing in Dorey, on the island of New Guinea, an epicenter of bird-of-paradise diversity. He frustratingly wrote:

> And this, too, in New Guinea! – a country which I might never visit again – a country which no naturalist had ever resided in before – a country which contained more strange and new and beautiful natural objects than any other part of the globe.[88]

His frustration was exacerbated by the isolation of the birds-of-paradise islands, which are located in the far east of Indonesia, closer to Australia than to the sophisticated cities of Jakarta and Surabaya.

Birds-of-paradise are found only east of what became known as the Wallace Line, the unmarked border Wallace identified that separates the flora and fauna of western Indonesia (tigers, elephants, rhinos, primates, wild cattle, hornbills) from that of eastern Indonesia (koalas, kangaroos, ground nesting birds, birds-of-paradise, cockatoos, parrots, cassowaries).

International Travel News

Paul Spencer Sochaczewski

Birds-of-paradise are important in traditional ceremonies. A Huli wigman from the Tari region of Papua New Guinea (left), and my friend Ida Pendana Gede Djelantik Putra Tembuku, a senior Balinese Hindu priest (right), use birds-of-paradise as ceremonial accessories and to perform rituals.

Frankie Fuller

Ruzaini Bin Ghazali, Lee Kong Chian Natural History Museum, Singapore

While glorious in the wild (see the videos produced by the Cornell Lab of Ornithology, the dead, cleaned birds-of-paradise stored in museum cabinets and examined by taxonomists lack vibrancy and give no indication of their flamboyant displays.

Two important examples of Wallace's standardwing (*Semioptera wallacii*) include:

Left: *Semioptera wallacii* was first collected in October 1858 on Batchian (now Bacan) island in eastern Indonesia by Ali, who is thought to have collected some 5,000 of Wallace's total haul of birds. This may be the same bird that Ali shot and proudly showed Wallace, noting "what a curious bird."

Above: Alfred Russel Wallace collected 8,050 bird specimens. He sold 5,000 bird skins but kept some 3,000 (representing about 1,000 different species) for his private collection. This standardwing specimen was identified as part of Wallace's private collection by his handwriting and the red stripe across the top of the label. It was collected in East Gilolo (now Halmahera) in 1859, possibly by Ali. It is now in the Lee Kong Chian Natural History Museum at the National University of Singapore. Wallace noted that this bird "differs a little from the Batchian [Bacan] specimens [shown above] in the much greater length of the breast plumes and other details." George Beccaloni, founder and director of the Alfred Russell Wallace Correspondence Project, says, "It took until 1881 for [Tommaso] Salvadori to describe the Halmahera bird as a new subspecies, which he called *halmaherae*. Shame he did not name it *alii* in honour of Ali."[183]

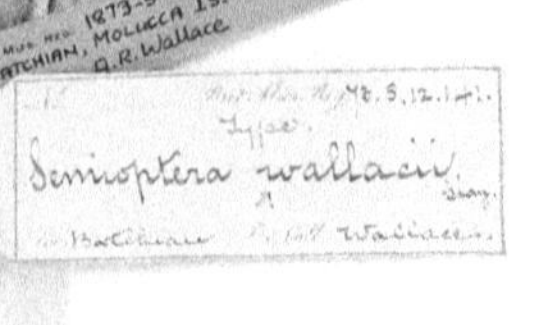

Jonathan Jackson, Trustees of the Natural History Museum, London.

And most of the bird-of-paradise species are limited to the huge island of New Guinea (the bigger-than-Texas-sized island is the second largest in the word, after Greenland) and surrounding smaller islands.

The terrain was unwelcoming:

> It seems as if Nature had taken precautions that these her choicest treasures should not be made too common, and thus be undervalued. The northern coast of New Guinea is exposed to the full swell of the Pacific Ocean, and is rugged and harborless. The country is all rocky and mountainous, covered everywhere with dense forests, offering in its swamps and precipices and serrated ridges an almost impassable barrier to the unknown interior; and the people are dangerous savages, in the very lowest stage of barbarism.[89]

Wallace rehired Charles Allen, who had been Wallace's first senior assistant, and asked him to undertake a solo collecting expedition to the island of New Guinea. Allen himself could have used an assistant like Ali. Despite having the assistance of the Dutch resident at Ternate and the sultan of Tidore, who sent a lieutenant and two soldiers to accompany him, his journey was a litany of obstacles, frustrations, and danger. Wallace wrote:

> To understand these [difficulties that Allen endured], it is necessary to consider that the birds of paradise are an article of commerce, and are the monopoly of the chiefs of the coast villages, who obtain them at a low rate from the mountaineers, and sell them to the Bugis traders. The natives are therefore very jealous of a stranger, especially a European, interfering in their trade, and above all of going into the interior to deal with the mountaineers themselves.
>
> [Allen] was told it was a three or four days' journey over swamps and mountains, that the mountaineers were savages and cannibals, who would certainly kill him; and lastly that not a man in the village could be found who dare go with him.

> The natives, refusing to obey the imperious order of the lieutenant, got out their knives and spears to attack him and his soldiers; and Mr. Allen himself was obliged to interfere to protect those who had come to guard him.
>
> [He] remained a month without any interpreter.[90]

Even when Allen made it into the interior, "only one additional species was found."

Alfred Russel Wallace had a complex view of the world, often acting and thinking in apparently contradictory directions. Most obviously, he had two very rational, very left-brained, very obvious reasons for wanting to collect birds-of-paradise. One, his only income came from selling specimens, and birds-of-paradise were sought-after items of great value to museums and collectors. And, two, they were of interest to scientists, and Wallace thought his studies of bird-of-paradise behavior, particularly his descriptions of their mating dances, could help him establish his scientific credentials with the intelligentsia in England.

But throughout his hero's journey in the Malay Archipelago, Wallace had a third reason for seeking these elusive birds: Several times in *The Malay Archipelago* he wrote about the beauty of the specimens he captured. Sometimes they were so exquisite, he was "in "a continual state of pleasurable excitement" for thinking about their natural beauty:

> In such a country, and among such a people, are found these wonderful productions of Nature, the birds of paradise, whose exquisite beauty of form and color and strange developments of plumage are calculated to excite the wonder and admiration of the most civilized and the most intellectual of mankind, and to furnish inexhaustible materials for study to the naturalist, and for speculation to the philosopher.[91]

SBS Eclectic Images, Alamy Stock Photo

Working through his agent in London, Wallace sold his bird-of-paradise skins to museums and individual collectors. But through his efforts he inadvertently contributed to the 19th-century fashion rage in Europe of wearing birds-of-paradise feathers and skins on hats that led to over-collection of the birds.

Malcolm Smith, author of *Hats: A Very UNnatural History*, wrote: "W. H. Hudson, an author, naturalist and leading British ornithologist of his day, recoiled with horror as he witnessed one sale of 80,000 parrot skins and 1,700 skins of birds of paradise late in 1897. 'Spread out in Trafalgar Square they would have covered a large proportion of that space with a grass-green carpet, flecked with vivid purple, rose and scarlet,' [Hudson] commented."

Today, all the estimated 41 species of birds-of-paradise[92] are protected through most of their range; their greatest threat now comes from destruction of their forest habitats. They are listed as Category II by CITES (Convention on International Trade in Endangered Species of Wild Fauna and Flora), which means trade in these birds is regulated. The populations of all but two of the 21 species are listed in the International Union for Conservation of Nature's Red List of Threatened Species.

But this book is about Ali, so the question I'd like to ask Ali is whether he, too, marveled at the beauty of these birds. Did he wonder why they have such almost-obscenely elaborate plumage, or why they dance in hyperactive avian jitterbugs that are unique to each species? Did he have qualms about killing them, or were they just a means to an end – part of his business agreement with Wallace – find 'em, shoot 'em, stuff 'em?

SECTION III

Where Did Ali Come From?

CHAPTER 8

The Pain of Changing an Idea Thought to Be True

While I am open to new experiences and ideas, I am semi-locked into my existing judgements and opinions. Call it human nature or a grumpy senior mindset, but it takes a fair bit of convincing to get me to change my mind about something.

It is difficult to change a long-held belief.

We sometimes resolutely stick to the idea that our belief is the only correct option and ignore opportunities to change our mind-set. Consider the psychological term "monkey trap," which refers to a simple contraption to catch a monkey. A banana or apple is placed inside a cage with a hole just large enough for the monkey to reach in and grab the big piece of fruit. Although the monkey could escape easily at any time by releasing its hold of the fruit, it refuses to give up its prize and is easily caught with its hand in the metaphorical cookie jar.

And so it was, with some resistance, that I was forced to back-track and revisit one of my core beliefs about Ali.

For years I believed (or, more accurately, chose to believe) the accepted wisdom that Ali was a "normal," uneducated, innocent lad from a village near Kuching, Sarawak, when Wallace singled him out and took the naïve teenager on a glorious boy's adventure.

Now, with new information, I'm forced to revise that certainty.

From Wallace's writings, we know that he hired Ali in Sarawak.

But he only vaguely mentions Ali's background — he refers to Ali as "my Bornean lad." Does that mean Ali was originally from the Kuching region, where Wallace was staying at the time? Or could Ali have had a more complicated backstory; specifically was Ali, in fact, from another part of Borneo?

Could he have been related to a noble Malay clan, even remotely?

And even more challenging to my accepted myth of Ali's origins, new circumstantial evidence indicates that Ali might have previously worked with — and travelled widely in the entourage of — a well-placed Englishman named Spenser St. John, who was Rajah James Brooke's friend, personal secretary, and biographer.

The result: My *preferred story has been questioned, and I'm forced to rethink the narrative.* It's as if a Shapeshifter, to use Joseph Campbell's term, jumped into the story arc and moved the goalposts.

So, with excitement, and a bit of trepidation, here are some new theories about Ali's background. As with much of what we call history, these theories are based on a fair amount of conjecture, speculation, and circumstantial evidence. Or more succinctly, as historian Jerry Drawhorn puts it, "we are left scratching for crumbs."

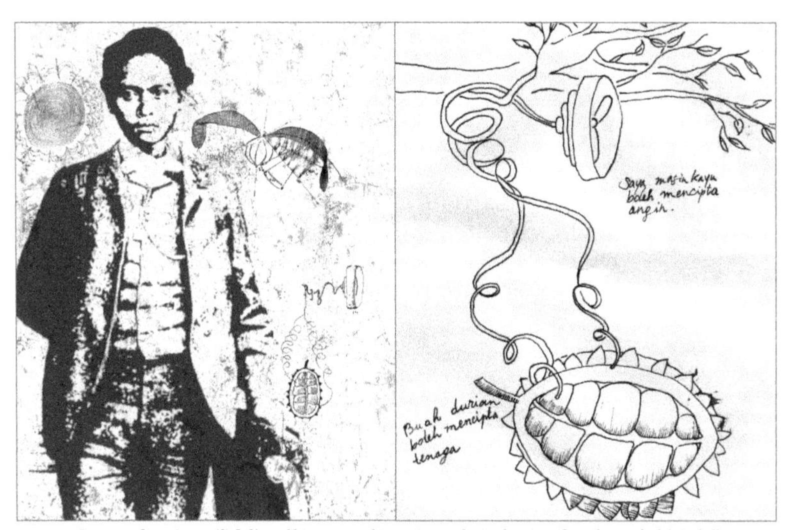

Portrait of imaginary Ali (left), and his purported invention to obtain electricity from durian (right), Isabelle Desjeux

Isabelle Desjeux, a Singapore-based artist, created a parody persona for Ali, positioning him as a nobleman and great scientist. Her name for Ali — Buang Bin Mohamed Ali — is a clever multi-lingual *jeu de mots*. Her tongue-in-cheek spoof (she calls it "speculative fiction," a disclaimer that appears in small print deep in the bowels of her website) includes a few slivers of truth and has received an inordinate amount of attention and reposting by gullible readers, particularly in Sarawak, evidence that it's rather easy to create a myth, particularly when dealing with a poorly known historical character. Proof that if you repeat a story often enough and slip in a few almost-facts, the fairy tale becomes truth. Reality is in the eyes of the reader.

Her tale draws on imaginative writers (particularly Tom McLaughlin and Suriani binti Sahari), supported by clever artwork (I particularly like her portrait of Ali as a well-dressed dandy). She tells of the discovery of a treasure chest in Ternate, which contains a manuscript putatively written by Ali. In her fable, unidentified researchers sent the manuscript to a Dr. Ola Modunagu, of the Department of Epistemology at the University of Failomics in Abuja, Nigeria, for review, resulting in his conclusion that Ali was a world-class scientist who, among other achievements, developed a way to generate large quantities of electricity from the odiferous durian fruit, years before Edison did so by more conventional means.

CHAPTER 9

Ali's Origin – Sarawak?

Let's examine the generally accepted claim that Ali was originally from Borneo.

Wallace described Ali as "my Bornean lad,"[93] and, in a 1867 lecture in London, referred to Ali as "a native of Borneo."[94] Wallace did not offer any additional details or clues, leaving the field open for speculation and educated guesses.

Most researchers, myself included, took the easy route (always dangerous) and believed that Wallace was implying that Ali was from the region then-termed Sarawak, specifically the area around Kuching, now the capital of the Malaysian state of Sarawak.

This was the assumption of John van Wyhe and Jerry M. Drawhorn, who, in 2015, confidently cited the certitude of conventional wisdom and wrote:

> It is likely that Ali was from the groups of Muslims living in various small villages of houses on stilts along the Sarawak River. He may also have come from the village of Santubong ... Ali was perhaps about 15 years old, dark, short of stature with black hair and brown eyes. He would have grown up on and around boats. He would have spoken the local dialect of Malay and was probably unable to read or write.[95]

This was the generally accepted "truth," an appealing legend of a young, inexperienced man plucked from a village and taken on a wondrous journey across Southeast Asia.

However, some eight years after his co-authored paper, Drawhorn, senior lecturer emeritus at Sacramento State University, suggests the possibility that while "Wallace certainly and repeatedly refers to him as 'my Borneo lad, did he mean 'Borneo Proper' [meaning the then-called town of Sarawak, now Kuching] or was he referring to Borneo as the entire island? There's ambiguity, and it's certainly possible that Ali was from another place [besides Kuching] in Borneo. Maybe Mukah, or Bintulu, or elsewhere."[96]

It is a classical historian's conundrum. The scholar might repeat a widely accepted scenario, then a rabbit hole appears, and then another, until we find ourselves in a maze of alternate options.

Imagined Conversation: First Encounter with a Strange Creature

We might imagine a scenario. Wallace is having dinner with his host, James Brooke, the famed White Rajah of Borneo. They are in Brooke's residence in what is now Kuching, the capital of Sarawak. Over wine and roast chicken, Wallace explains his plans to spend a few years exploring the isolated islands of Indonesia. The dialogue might have gone like this:

"And you're going on this trip by yourself?" Brooke asks, pouring Wallace some more wine.

"As I've done in Brazil on the Rio Negro."

And how did that work out in the end? Brooke thinks. Instead he inquires, "And you speak Malay fluently?"

"Er, not yet."

"And you have a diplomatic right of passage and introductions?"

"You know I don't."

"And you have some form of income?"

Wallace doesn't bother to answer.

Brooke leans back in his chair. "Wallace, you're a good chap. But they're going to eat you alive down there in Indonesia. Here, you're my guest, so the shopkeepers don't cheat you too much, and the bandits leave you alone. But down there . . ."

Brooke calls out for the Malay man who runs Brooke's household. We don't know his name; let's call him Abdul. "Abdul," Brooke says. "You've got relatives all over the place. Find my dear, innocent friend Tuan Wallace a young boy to accompany him on his travels, to cook for him, to teach him Malay, and to see that he doesn't get into too much trouble."

A few days later Abdul presents a quiet, short boy of 15. "This is my cousin Ali. He's not a bad cook. He's from Santubong, where Tuan Wallace is going to work."

Brooke asks the young man, in near-perfect Malay but with a lilting English accent that almost makes Ali laugh, whether he is prepared to accompany this strange, gawky white man to Indonesia on a journey from which neither of them might return.

Ali lowers his eyes. He had never been spoken to directly by a white man, let alone the rajah. Indeed, the man identified as Wallace is a strange creature. He is too tall. He wears odd clothes. His eyes are intense. But he has a kind, expressive face. Ali, who had never left the village he came from, who only vaguely knew where Indonesia was (to the south, basically), who understood just a few words of English, who . . . well, you get the picture . . . simply says softly: "Yes, Tuan."

"He'll do," Wallace says, clapping his hands.

How and where did Wallace and Ali meet?

Here the speculative juices run freely.

I had originally thought it was likely that it was Rajah James Brooke who offered Ali's services to Wallace, thinking that Ali might have been a member of Brooke's household or perhaps a relative of one of his local staff (see "Imagined Conversation: First Encounter with a Strange Creature").

But Drawhorn suggests there might be another explanation of how Wallace met Ali.[97] His argument is based on the writing of Spenser St. John, Rajah James Brooke's personal secretary and later British consul general to Labuan and Brunei. In his book *Life in the Forests of the Far East*, St. John mentions a servant boy/cook named Ali. St. John notes Ali's cooking skill during a trip to Brunei and northern Borneo, when St. John was investigating the death of the merchant Robert Burns by pirates:

> [Tuanku Yasin] furnished us with supper, cooked by my servant, Ali; omelettes, stews, sliced sweet potatoes, rice, soup, which we enjoyed, and a bottle of wine made the meal complete.[98]

The event took place in 1851, which, if it truly refers to Wallace's Ali, would have been when the young man was roughly 11 years old.

And it occurred in the pirate-infested Sulu Archipelago situated between Sabah, Malaysia, and southern Philippines. So, the Ali who worked for St. John certainly had a taste for distant travel and working under difficult conditions before signing up with Wallace.

The circumstances, recorded in Wallace's and St. John's journals, and brought to my attention by Drawhorn, suggest it is possible that Ali moved from St. John's entourage to Wallace's in late 1855. That leaves the questions of how and why such an exchange took place?

We know that St. John accompanied James Brooke and Wallace to Peninjauh (now Peninjau), outside Kuching, where Wallace collected a huge number of moths. It is possible that Ali, as St. John's servant, was also present, and Ali and Wallace could have met at that time.

The final, perhaps conclusive, clue is that *Ali completely disappears from St. John's narrative* following the Peninjauh visit. From 1866, St. John instead writes about an apparently new cook and servant, a

Chinese lad named Ahtan.[99]

It's not a smoking gun, but these clues suggest that Ali was in the employ of St. John before being engaged by Wallace, and the transfer of employment took place in late 1855, either in Peninjauh or at Christmas celebrations at James Brooke's residence in Kuching.

But while this job change might well have occurred, one big question remains: St. John was apparently satisfied with Ali, and, as they say, "It's hard to get good help these days." So why would he have "traded" Ali for Ahtan?

Here's what *might* have happened in Rabbit Hole #39083. Drawhorn writes:[100]

"Spenser St. John's consort was a Malay woman named Dayang Kamariah, who was from the *perabangan* class of aristocratic Malay families, which took refuge in Sarawak during the interclan wars and successional disputes in Brunei during the early 1840s - 1850s. After the assassination of Muda Hassim and his brothers, their families and the few surviving princes who were in opposition to Sultan Omar Ali fled to Kuching. Ali might have been a member of Kamariah's household staff or possibly even a member of her family. It was dangerous for Ali to reside with St. John in Brunei in his new posting as consul, since Kamariah's family supported the Rajah Muda Hassim line of succession in Brunei. St. John would have needed a new personal servant and cook [and] St. John may have been amenable to allowing Ali to become Wallace's servant."

This speculative scenario is derived from the white colonial perspectives of Spenser St. John and Alfred Russel Wallace: *Ali was a servant, a mere cook. His feelings and desires didn't matter too much to his employers. Or did they? I'm trying to imagine Ali as a modern-day young professional baseball player under contract to a world-class team, let's say the New York Yankees. One hot day in July, the manager calls the player into his office after the game and tells him: "Sorry kid, you've been traded." The kid is speechless. Playing for the Yankees was his childhood dream, and sure, in his head he understands the way the baseball business works, but still… "To where?" "Cincinnati. Pack your bags. You leave tomorrow." The player wants to cry, but, of course, he doesn't. He tries to protest but knows it's futile. The manager chews on an unlit cigar. "Sorry kid. Good luck."*

Is that similar to the way Ali was treated? Like a helpless pawn without a say in his future? Spenser St. John was close friends with White Rajah James Brooke. St. John was being appointed as British consul to Brunei. He was the New York Yankees, he was Real Madrid, he was, well you get the idea. But this guy Wallace? He was the Cincinnati Reds — a mediocre team in a nothing city that hadn't tasted importance for decades. Did Ali go quietly with a touch of his forelock? Did he go with vinegar stirring in his soul? Did he curse his destiny? Or was St. John, in spite of his high position, a bastard to work with, and Ali was pleased to see his back?

Among the few people in Sarawak who care deeply about the Ali saga are Tom McLaughlin, an American historian, and his Sarawak-born wife, Suriani binti Sahari.

They offer a contrary view about Ali's origin, which has been widely discussed (accepted by some, disdained by many) in Kuching. They base most of their case on interviews with elders living in Malay communities around Kuching and, notably, the pronouncements of a prominent *bomoh*, or Malay shaman, named Sapian bin Morani.

These informants claim that Ali came from an upper-class, exceptionally well-placed family; his patronym was Ali bin Amit (Ali, son of Amit); he was born around 1849; he was the youngest of five children; and he came from Kampung Jaie, today about two hours' drive from the Sarawak capital of Kuching. McLaughlin and Sahari suggest that Ali worked in Rajah James Brooke's household under the tutelage of his elder brother Osman, later known as Panglima (General) Seman, and subsequently moved to Osman's residence in Kampung Panglima Seman. They say that Ali was influenced by complicated palace intrigue, and "Ali probably followed his brother through battle, helping to bandage wounds and cooking for the troops ... Ali knew that his fortunes lay with the white people who came to rule Sarawak. He befriended a person named 'Edward' to learn the English language. He was no stranger around the Astanna [*sic*] where the English Rajah lived."[101]

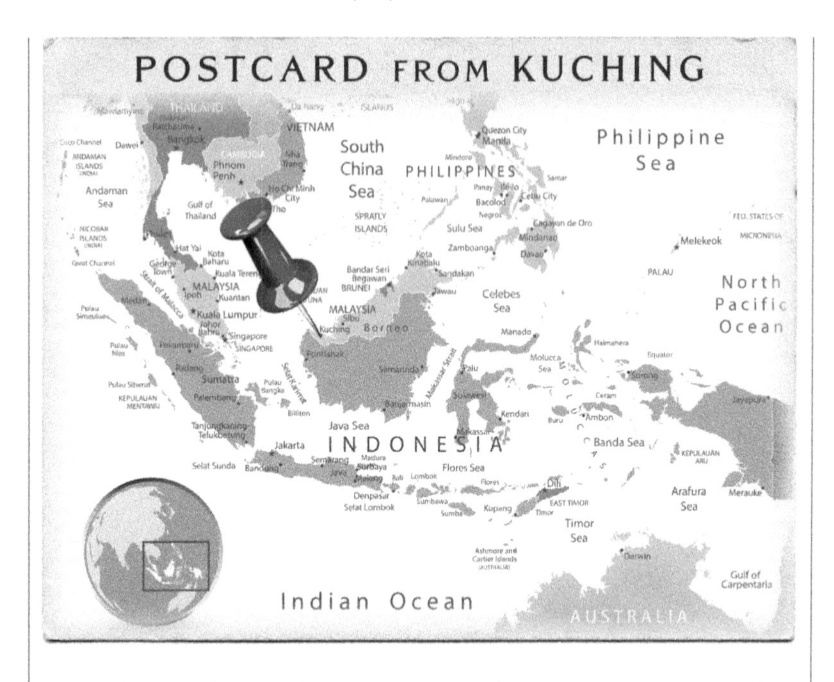

Right after graduating from George Washington University in 1969, I enrolled in the U.S. Peace Corps and was posted to Sarawak, one of two Malaysian states on the island of Borneo. It was here, while visiting isolated primary schools that served community longhouses, that I was introduced to several themes that have guided my life — nature conservation, tropical ecology, changing cultural patterns, and the life and achievements of Alfred Russel Wallace.

For me, Kuching remains one of the most livable, interesting, and friendly small cities (pop 570,000) in Asia. It also has, arguably, since such things are subject to never-ending debate, the most intriguing food culture in Asia.

The first White Rajah, James Brooke, invited Wallace to Sarawak; that sojourn changed Wallace's life and the life of Ali. Similarly, my experiences living for two years in Sarawak, on the long and vital Baram River in the north of the state, changed my life by exposing me to rural tribal cultures living in a rapidly changing society.

I return to Sarawak every year or so, and before heading upriver, I stay in a comfortable hotel in Kuching that offers a majestic view of the Sarawak River with Mount Santubong, where Wallace wrote his famous Sarawak Law, in the distance. I take a sampan across the river to Fort Margherita, an imposing whitewashed complex that was built by Charles Brooke, James Brooke's nephew and the second White Rajah of Borneo, then admire the nearby gardens of the Astana, originally Charles Brooke's residence, now the residence of the state governor.

Wikipedia

The Sarawak Museum, created in 1860 by the second White Rajah, Charles Brooke, at the suggestion of Alfred Russel Wallace.

Like most visitors to Kuching I visit the Sarawak Museum, one of the world's more interesting museums. And Wallace had a hand in its creation — he suggested to Charles Brooke that such a remarkable place as Sarawak deserved an institution to educate people about the area's rich cultural and biological diversity.[102]

But there is another aspect of Sarawak that has challenged me more than Wallace, friends, food, or golf (one of my favorite courses in the world is at Borneo Highlands, outside Kuching): the destruction of Sarawak's tropical rainforest and the corresponding loss of tribal land owned by indigenous people and the devastation of the state's biodiversity, including the emblematic orangutan that so fascinated Wallace. First comes deforestation, then oil palm plantations. Linked to this wanton example of ego, greed, and disregard for basic human and ecological rights are the bigger-than-life stories of several notoriously corrupt and uncaring people in high positions, as well as a few genuine heroes, such as Bruno Manser and Peter Kallang. (More on these individuals and their fictional counterparts can be seen in my novels, *Redheads* and *EarthLove,* and non-fiction books, including *An Inordinate Fondness for Beetles* and *A Conservation Notebook.*)

Paul Kingsley, Alamy Stock Photo

The Kenyah "Tree of Life" in the Sarawak Museum, painted by my friend, the late Tusau Padan.

Shaiful Zamri Masri, Dreamstime.com

Mount Santubong, downriver from the city of Kuching, Sarawak.

CHAPTER 10

Ali's Origin – Ternate?

Two well-regarded historians, the Earl of Cranbrook and Adrian Marshall, suggest that Sarawak has *not* been confirmed as Ali's birthplace and offer various supporting arguments.

They offer circumstantial evidence that Ali was "a roving youth of Malay race born and raised outside Sarawak, and it is plausible that Ali's 'own country' was Ternate, where he had grown up hearing the Dutch style of Malay, where he found his wife with surprising alacrity and where he chose to settle permanently."[103]

They argue that "Ali's domestic abilities – 'he was clean and could cook very well' – denote independence from his family and possibly some previous experience as a servant. His evident capability to organise his employer's affairs, with a level of command over other men, implies maturity and self-reliance. Ali's language, his skills and his self-confident authority are wholly inconsistent with the image of an unsophisticated Sarawak lad . . . Ali's prowess as a boatman could have been gained through serving on inter-island voyages."[104]

Cranbrook and Marshall note that in one passage in *The Malay Archipelago* Wallace quotes Ali; this is one of only two times we hear a direct quote attributed by Wallace to the young man (the other time, reported in English translation, is when Ali shows

Wallace the "curious bird" he had just caught; that phrase forms the title of this book).

Wallace wrote that he and Ali were staying in a house on Arru (now Aru) island in 1858:

> [The house contained] about four or five families & there are generally from 6 to a dozen visitors besides. They keep up a continual row from morning to night, talking laughing shouting without intermission ... My boy Ali says "Banyak quot bitchara orang Arru" [roughly *The Aru people are very strong/loud talkers*] having never been accustomed to such eloquence either in his own or any other Malay country we have visited.[105]

A small number of historians (maybe enough to form a football team) have expended considerable energy dissecting this phrase, considering it a cipher that will unlock the mystery of Ali's origin.[106]

So far, this debate has not devolved into social media fisticuffs, as I have witnessed in other academic disputes. Cranbrook and Marshall say that the phrase "is far from the vernacular of Sarawak Malay [which is a distinctive dialect] and shows that, early in his employment, Ali addressed [Wallace] in the Bazaar Malay typical of the area of Dutch control."[107]

I disagree with Cranbrook and Marshall's conclusion that the "Banyak quot bitchara orang Arru" quote indicates Ali was not from Sarawak.

First, the incident occurred after Ali had been with Wallace for two years; plenty of time for him to have become accustomed to speaking "standard" Malay, also called Straits or Singapore Malay, which was the lingua franca of the region and the form of Malay that Wallace most likely used himself.[108]

Second, Wallace was recalling a relatively unimportant conversation, and there is no reason to think that Wallace would have remembered it precisely (can you remember verbatim a mundane conversation you had just yesterday?).

Third, Wallace provided the *essence* of Ali's statement, using the standard Malay that would be recognizable to any reader familiar with the region. Writers "approximate" all the time. I don't place too much credence that this particular argument answers the questions about Ali's origin.

There are other questions about Cranbrook and Marshall's claim that Ali originated in Ternate. They argue that the "surprising alacrity" by which Ali found a wife in Ternate would only have been possible if he had originated from that island. Jerry Drawhorn offers a rebuttal: "[Despite decimation of the nutmeg and clove gardens by the Dutch] there were still opportunities for [Ternate] families to become involved in trade. Ali might have appeared a promising financial asset as an in-law to Ternate families."

Drawhorn observes "the claim by Cranbrook and Marshall that Ali was a displaced Ternate immigrant is curious. A review of [Rajah James] Brooke's writing and those of contemporary authors ... indicate ... there is no evidence of Ternate immigrants ever having been established [in Sarawak] during the James Brooke period."[109]

Additionally, Wallace was careful to describe his assistants by their geographical origin or ethnicity – "Lahagi, a native of Ternate," "Lahi, a native of Gilolo," "Loisa, a Javanese cook," "Manuel, a Portuguese of Malacca," "[Inchi Daud,] an Amboynese Malay."

He always refers to Ali as being "my Bornean lad." This doesn't necessarily mean that Ali came from the town we know as Kuching; it could simply mean that he came from the island of Borneo. Wallace and Ali spent some six years together, and Ali must have told Wallace about his upbringing, otherwise Wallace, who was a conscientious writer, would not have written repeatedly about Ali's Borneo origins.

I tend to agree with Jerry Drawhorn that "there is no plausible evidence that Ali was anything other than what Wallace claimed he was, a Bornean Malay. It is most likely he came from where Wallace recruited him, Sarawak." [110]

SECTION IV

Where Did Ali "Retire"?

CHAPTER 11

Ali: Rich, Facing New Adventures

In early 1862, after years travelling together throughout the Indonesian archipelago, Wallace and Ali made their way to Singapore. Wallace was getting on a ship to return to England, but first he took Ali to a photographer; the picture (the only one we have) shows a full-mouthed, serious, dark-complexioned lad with wavy hair, thick eyebrows, and a broad nose, dressed in a dark European-style jacket and under-jacket, white shirt, and a white bow tie. Was Wallace being sentimental and desirous of having a personal memento of Ali? Or was he being generous and giving Ali a great gift — a photo portrait of him (taken by a European photographer, just like the Malay sultans and rich Chinese traders) during the early days of photography?

Wallace remembers the farewell:

> On parting, besides a present in money, I gave him my two double-barrelled guns and whatever ammunition I had, with a lot of surplus stores, tools, and sundries, which made him quite rich. He here, for the first time, adopted European clothes, which did not suit him nearly so well as his native dress, and thus clad a friend took a very good photograph of him. I therefore now present his likeness to my readers as that of the best native servant I ever had, and the faithful companion of almost all my journeyings among

the islands of the far East.[111]

Collection of the National Museum of Singapore, National Heritage Board

An 1860 photograph of Johnston's Pier (located near the current Collyer Quay at Marina Bay) from where, on February 20, 1862, Wallace would have likely bid farewell to Ali. From there, where did Ali go? Did he return to Sarawak? Or to his wife and family in Ternate?

When Ali parted ways with Wallace, where did he "retire?"

We aren't sure. We have ideas and snippets of evidence, a bunch of clues and a jumble of hunches. But we don't have a factual smoking gun. Or do we? The search goes on.[112]

Ali was about 21 or 22 when Wallace left. Let's consider Ali's thought process. What decisions did he make? What choices did he have?

When you were 22 did you have a clear career path lined up? Maybe you did, and you had a hot contract in your hand to work as a poorly paid intern at Merrill Lynch, a legal document that came with the distant promise of a lucrative career in finance and the immediate thrill of living in one of the outer boroughs of New York City and sharing an apartment with three other ambitious young professionals. But most people, I'm willing to guess, don't have a career path clearly sign-posted at that age.

Remember the adage: "If you want to make god laugh, tell her your plans." Ali knew he should return to Ternate to be with his wife and children. But did he want to? Did he enjoy the feeling of independence? Yet he needed to work somewhere, and the work he was best at was specialized but potentially rewarding — caring for rich, white people and shooting and skinning birds.

Did Ali have enough self-confidence to go out on his own? Did the call of his homeland — Borneo — have enough gravitational pull for him to remain? Did he acquire the "wanderer" gene that he had absorbed from travelling with Wallace?

Or to put it in a modern-day film context, did Ali wish, with all his heart, that the scene had worked out like this: The day before his departure Wallace had given Ali guns, supplies, and cash as a farewell present. The following day, Wallace and Ali are standing at Johnston's Pier. Wallace is about to board a lighter that will take him to the steamship anchored in the harbor that is heading toward points west. The whistle to board sounds. Wallace takes a step forward but turns to Ali and says: "Ali. I can't leave you here. Want to go to England?" And Ali would have said . . .

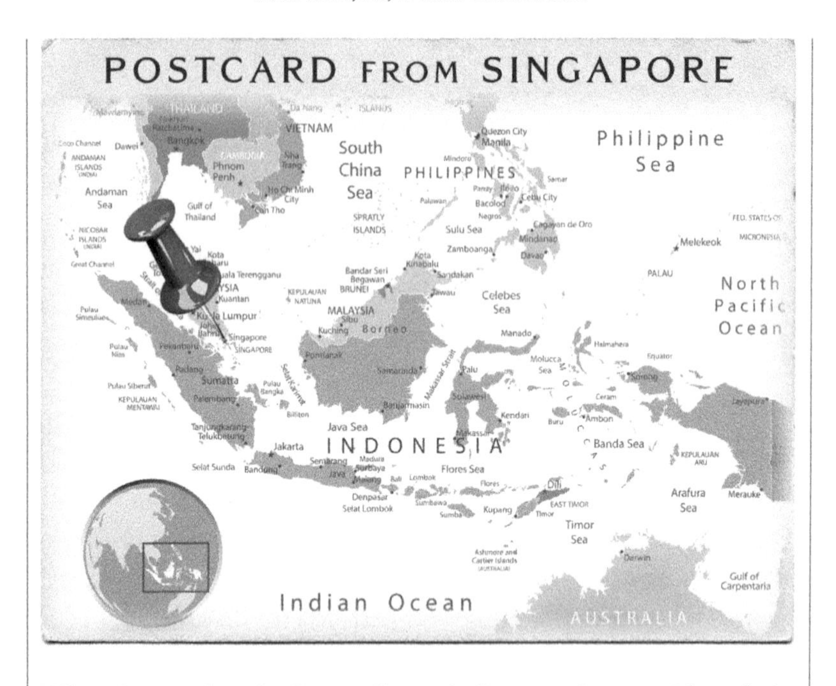

When I served in the Peace Corps in Borneo, I was paid a subsistence allowance of $50 a month. Once a year I saved my pennies (and dug into an emergency reserve my parents had given me) to catch a freighter from Kuching to Singapore for a bit of R&R (years later I tried to recreate those two-day voyages but was told that the only way I could take a ship to Singapore would be inside a shipping container). On arrival I would check into a cheap, noisy Chinese hotel on Beach Road and scour Change Alley for an Indian money-changer who might give me a few extra cents on the dollar for my precious bank notes.

Singapore was where Wallace landed in 1854 and where he departed in 1862. He was fascinated by the multi-ethnic, multi-lingual hustle of the place, an energy that Singapore has built on to become one of the most successful small city-states in the world. He wrote: "Few places are more interesting to a traveller from Europe than the town and island of Singapore."

It seems that whatever Singapore sets out to do, it strives to do it better than any other country. A partial list of its top-ranked achievements: harbor, national airline, airport, public transportation, cleanliness, safety, tech research, universities, zoo, botanic garden, and, in recent years, an exceptional commitment to creating a sustainable, green urban environment. It has four official languages. People get along. They are rarely frivolous. They do not jaywalk. They are world-class shoppers, and even better eaters.

Shutterstock

Singapore, seen from the summit of Bukit Timah. While much of the country's nature has disappeared due to rampant urbanization, pockets of wilderness remain, where new species are regularly being discovered. Singapore is home to more than 40,000 wild, native, non-microbial species, while the number of plant species growing in Singapore's Bukit Timah Nature Reserve is more than that in the whole of North America.

As I was nearing the end of my Peace Corps stint, I decided I wanted to stick around Asia for a while. I saw an ad in *The Straits Times* for a job with an advertising agency in Singapore. The owner of the small regional agency, Roland Tan, gave me an assignment during the interview – come up with an advertising campaign for something, I forget what. Although I had zero advertising experience, I had worked in journalism, edited the university newspaper, and handled public relations for the Merriweather Post Pavilion, a summer music venue outside Washington, DC. (the highlight of which was when I delivered two bottles of Southern Comfort to Janis Joplin's dressing room).

I spent a few hours in the quiet of a local library and the next day presented Roland with my ideas. He hired me, arranged for a work permit, and gave me a low salary with accommodation in an annex off the office. I needed a job; he needed a white guy who would work hard and cheap. Thus started my advertising career, which "put rice on the table," for more than a decade.

Paul Spencer Sochaczewski

Bukit Timah, at 538 feet, is as high as mountains get in Singapore. But this hill, now a nature reserve, is where Wallace collected some 700 species of beetles.

Even when this occupation took me to live in Jakarta, Indonesia, I returned to Singapore every few weeks. Indonesia, at that time, had a nascent advertising industry, and I had no choice but to go to Singapore to work with reliable art directors, commercial film producers, and printers. And in Singapore resided my friend Horace Wee, a musical genius (he played five instruments) who helped me create memorable jingles. (I think all advertising works better with a catchy jingle.) I'd walk in to his studio (where I learned how empty cardboard egg cartons can be used as sound-proofing), blurt out a semblance of a melody and a few lyrics, and he'd take it from there. "Hey, let's try a bubble-gum pop version," he might say, strumming his guitar. "Or calypso? Romantic ballad?" I owe much of my modest advertising success to Horace. The highlight of my creative vision: the lyrics and jingle for a TV commercial promoting a mosquito coil made by a client in Indonesia. Translated from Indonesian, the ditty boasted: "Kills mosquitoes 'til they're dead."

CHAPTER 12

Ali Returned to Sarawak?

Malaysia has two states on the island of Borneo – Sabah and Sarawak. I have a particular fondness for Sarawak, since that is where, in 1969, at the age of 22, I began two years working as an education adviser for the U.S. Peace Corps. Although the state has become more sophisticated in the years since, it retains a casual semi-rural vibe enhanced by strong cultural traditions. It is a good example of an ethnic, linguistic, and religious melting pot, with residents from all the major faiths and some dozen or so ethnicities, including various tribal groups, all living in relative harmony. Also, like much of the rest of Southeast Asia, Sarawak has diverse and fascinating natural diversity, which is being hammered by oil palm plantations encouraged by the toxic cocktail of corrupt politicians, greedy businessmen, and apathetic consumers.[113]

People of Sarawak have a modest awareness of Wallace and little awareness of Ali.

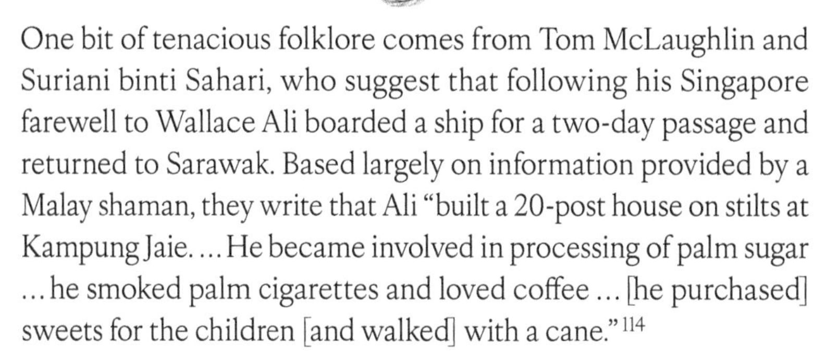

One bit of tenacious folklore comes from Tom McLaughlin and Suriani binti Sahari, who suggest that following his Singapore farewell to Wallace Ali boarded a ship for a two-day passage and returned to Sarawak. Based largely on information provided by a Malay shaman, they write that Ali "built a 20-post house on stilts at Kampung Jaie. ... He became involved in processing of palm sugar ... he smoked palm cigarettes and loved coffee ... [he purchased] sweets for the children [and walked] with a cane."[114]

McLaughlin and Suriani were told Ali had a black sea chest of British origin, which they say contained a "picture of Ali with a European gentleman [and] papers with British seals." The box was evidently haunted because it "became alive each Thursday night making noises. Because of the belief in ghosts, the box was taken out and placed in the sea."

For some two years, from May 1, 1859, to May 4, 1861, while Wallace was travelling in Indonesia, Ali seemingly disappears from Wallace's chronicle. Some observers call this Ali's Gap Years. Sarawak-based fabulists McLaughlin and Sahari believe that during this period Ali "probably returned to Kampung Jaie to check on his nephews" and to return for the Hari Raya (Eid) celebrations. Jerry Drawhorn disproves the theory that Ali returned to Sarawak, citing numerous entries in Wallace's journals that during this period Wallace was in regular contact with Ali, who remained in his employment as an independent contractor.[115 116]

McLaughlin and Sahari also state that according to their informants, after Ali returned to Sarawak he married a woman named Saaidah binti Jaludin.[117] And they quote a long *pantun* (a form of traditional song often used to relate myth and historical events) sung to them by Jompot bin Chong, a descendant of Panglima Osman (Seman) who they say was Ali's elder brother. Verses include statements like these (in translation from Sarawak Malay): "How are you, Wallace the white man/Wallace and brother Ali are good friends/Unfortunately this year our team work has ended." McLaughlin told me he was skeptical with the bomoh's observations until Jompot came up with the *pantun*. "He didn't even have to think about it," McLaughlin told me. "He just spouted it out. Then I checked with an 86-year-old '*pantun* expert' here in Kuching, and she said she had heard of it. The *pantun*, in itself, is very strong evidence that Ali was here."[118]

And finally, they were told, and propagate the suggestion, that "[Ali] died just after the Japanese invaded Kuching," which would have made him about a hundred years old at the time of his death.

They visited what they were told was Ali's grave in Kampung Jaie. In an attempt to "test" the authenticity that this was indeed Ali's grave, McLaughlin and Sahari brought a bomoh and a local official to the site. They said that the government official was so overpowered by the spirit of Ali that he collapsed on top of the grave.

Frederick Boyle and his brother, Arthur, travelled in Sarawak in 1863 and hired a young man they named Ali Kasut (Ali of the Shoes) as their competent and reliable boat captain and camp manager. It is likely that he was the same Ali who worked with Wallace, but this is subject to confirmation. The circumstantial evidence, as noted in Frederick Boyle's book *Adventures Among the Dyaks of Borneo*: "Ali Kasut was a skilled guide, sailor, and camp manager. The timing of his employment with the Boyle brothers fits. He was reliable and well-versed in Sarawak customs and culture, spoke some English, and, tellingly, wore English clothing and black leather shoes instead of traditional Malay attire."

And then there is the curious story of Ali Kasut ("Ali of the Shoes"), a saga first brought to light by Jerry Drawhorn, a retired American historian and biological anthropologist based in Sarawak. He believes "Ali Kasut" was almost certainly Wallace's Ali.[119 120]

The circumstantial evidence is, in my mind, convincing, but we must be clear that Ali Kasut has *not* been confirmed as the Ali who accompanied Wallace.

The tale of Ali Kasut was told by a young English adventurer named Frederick Boyle, who wrote two relatively obscure books mentioning the young man: *Adventures Among the Dyaks of Borneo* (1863) and *The Savage Life* (1876).

Boyle's nickname of this Ali as Ali Kasut (Ali of the Shoes) is based on the black leather shoes the man wore, explained a bit later.

In his adventure-memoir, *The Savage Life*, Boyle writes of spending June 1863 in Changhi (now Changi, where the airport is located) on the island of Singapore:

> It was in June, 1863, that my brother and I hired Mr. Reid's bungalow for a month's picnic.[121]

On subsequent pages he mentions Ali Kasut by name four times, establishing that Frederick Boyle (and his brother Arthur) had met and hired Ali Kasut in Singapore, in or before June 1863. This timeline would suggest that after Ali saw off Wallace in February 1862, he stayed in Singapore for a year and several months.

Ali Kasut was knowledgeable about the customs and politics of the royal families of Johor (now the Malaysian state of Johor), just across the straits from Singapore:

> On a slope near by stood a very large house, and another somewhat above. Flags were displayed on each of them, and armed men hung round. Very much astonished, I turned back, and questioned Ali Kasut — Ali of the Shoes — our guide, interpreter, and friend. "That's the palace of the Datu Tumangong's mother," said he. "The Datu himself has a house here." [122]

Boyle recounts Ali's observation about a conversation he had with the Tumangong:

> The Tumangong said no word about our voyage, but Ali Kasut told us afterwards, that everyone knew how we had reached Changhi.[123]

Boyle admires the furnishings in the Tumangong's house. The exchange shows that Boyle respects Ali's opinion, and Ali was close enough to Boyle to reply "scornfully" and express his opinion without fear of reprisal:

> I asked Ali Kasut if these [curtains] were of English manufacture, and scornfully he replied. Nothing European will a Malay use, in weapons or clothes, if he can buy the vastly more expensive products of his own land — and he is right.[124]

Finally, at one point Boyle is engaged in a delicate conversation with the Tumangong concerning the territorial intentions of Rajah James Brooke and the likely response of England to the Tumangong's planned military expansion of his own domain.[125] Boyle implies that Ali Kasut saved him from getting stuck in a tricky discussion that had no easy answers:

> The chief said nothing. Roused by these suspicious inquiries, we asked what force the Tumangong possessed? No answer came, and presently Ali Kasut entered, crouching, to offer us His Excellency's own bath. [The Tumangong] smiled and motioned us away.[126]

The brothers then travelled to Sarawak, apparently guided by Ali Kasut.[127]

Frederick Boyle was not the careful diarist that Wallace was, and only occasionally mentions dates. But as they prepared to leave Sarawak, Boyle specifically refers to "our experiences of four months," indicating that they stayed in Sarawak roughly between July to October 1863, immediately after travelling with Ali Kasut in Singapore.[128]

Alfred Russel Wallace mentioned Ali by name some 42 times in *The Malay Archipelago* and refers to him as his "faithful companion," evidence of the importance the young man played in his travels.

Similarly, Boyle mentions Ali Kasut by name 14 times in *Adventures Among the Dyaks of Borneo* and refers to him as "friend."

As with other Ali-related theories, we must rely on clues. And several clues that Boyle offers are significant.

Ali Kasut was a loyal, competent, and reliable camp manager, boatman, and assistant, characteristics that were very similar to those exhibited by Wallace's Ali:[129]

> Ali Kasut, our guide, interpreter, and friend, soon procured us a large sampan suitable for the purpose ...Arthur and I lay under the "kajang" or matted roof in the centre, our boatmen droned out nasal songs in the bows, and the trust Ali squatted behind, steering our course with unerring paddle.[130]

And they relied on Ali's judgement:

> We prepared to return, but the sky looked so black to the eastward as to make us hesitate. Trusting, however, to the wisdom of Ali Kasut, which decided that we could easily run across before the squall broke, we embarked in the sampan, and hoisted sail . . . about half way to shore the squall came down . . . the mats were whisked through the air into Ali's face as he sat behind steering our course . . . the waves rose in a moment, a mist surrounded us so dense that we could not see a boat's length ahead; the crew squatted in the bows, silent, drenched, and helpless, while the trusty Ali, assisted by Arthur's servant, still kept the head of our sampan to the waves . . . under the lee of a huge rock Ali beached the sampan, and once more we stepped on land . . . the experienced Ali himself admitted to us that he had not anticipated the superintendence of our affairs again, and regarded our return as a striking instance of the power of fate.[131]

But he wasn't perfect:

> In the morning Ali Kasut despatched some of our crew to Sabooyong with hooks and provisions, but he failed to send us our sketching materials.[132]

As might have been expected by good colonials, the Boyles complained frequently about the quality of servants, and spoke, again, about how happy they were with Ali Kasut as their "general manager":

> All who have travelled in the East will have taken for granted that the carelessness and stupidity of our servants were the greatest annoyance we had to endure, and their own experience will suggest the details which I have spared. Those who as yet are acquainted only with the harmless delinquencies of Jeames or Syusan, can form no idea of the vexation an Asiatic servant can cause, without a long catalogue of trivial grievances and annoyances, which we now endeavour to forget in the pleasant recollection of our travels. We were indeed fortunate in securing Ali Kasut as our general manager.[133]

And Ali Kasut was often ill — we know Wallace's Ali also suffered poor health, he was seriously ill no fewer than six times on his journey with Wallace (Chapter 4, "Ali's Career Path — In Sickness and in Health"):

> [Ali Kasut's] health was very delicate, and he was continually invalided.[134]

The almost-smoking gun lies in Ali's sartorial preferences, which resonate with Ali's decision to be photographed in Singapore wearing European clothes rather than traditional Malay dress:

> "Kasut" signifies shoes. Ali was distinguished from his innumerable namesakes by the practice of wearing such articles, which are not commonly affected by Orientals. Ali possessed two or three pair made of black English leather, without which he was never visible.[135]

When the Boyles left Sarawak they gave Ali Kasut significant presents, as did Wallace when he left Singapore:

> After my brother's return from Seribas, we began to make our preparations for departure. The large sampan, or boat, we presented with all its fittings to Ali Kasut, who had discharged the duties of his responsible station as interpreter and general guardian of our interests in a manner worthy his high reputation as protector of the griffins. His complexion of brownish-green fairly became unctuous with delight when he realized that the magnificent craft he had steered so often from island to island beneath the green peak of Santubong, and through the hot, silent river-reaches under shadow of vast forest trees, or between quivering belts of mangrove and nipa-palm, was actually and absolutely his property.[136]

But Boyle goes a step further than Wallace and speculates on what might have gone through the young man's imagination.[137] One might argue that their generous gift might have induced Ali to stay in Sarawak. He had a valuable boat and a deep knowledge of Sarawak geography, customs, and trade practices. He was also a good manager, spoke enough English to set himself apart from other local traders, and had useful contacts with both local and foreign residents. Boyle writes:

> Visions of trade with the simple Dyaks at a profit of 1000 per cent, formed themselves into a dazzling picture before his eyes; a hundred Malay pleasures became tangible in the near future, and possibly the wealth and influence of a real Nikodah [*nakhoda*, from

> the Persian, ship captain, a rich merchant owning a merchant vessel] showed themselves at the end of the vista, with, maybe, half a dozen honestly purchased wives, and a steeple-crowned residence like that of the uxorious merchant at Muka.[138]

The idea that Ali Kasut was the same young man who worked with Alfred Russel Wallace "feels" plausible.

But, as with other tantalizing tidbits of slippery history, this particular speculation is open to question. Frederick Boyle does not tell us anything about Ali Kasut's background. It is, of course, possible that a search in dusty library archives by some curious scholar might unearth Boyle's diaries or notes, at which time a bookish Philip Marlowe emerges shouting "I know the secret!"

Drawhorn notes that "Frederick Boyle never mentions [in *Adventures Among the Dyaks of Borneo*, published in 1865] that 'Ali Kasut' was 'Wallace's Ali,' but then again Wallace had not yet become the famous writer of the 'Malay Archipelago [published in 1869].'"[139] [140] Nevertheless, one might point out that Boyle wrote about his Singapore adventures in *The Savage Life*, published in 1876, which was years after Wallace had returned to England, had written *The Malay Archipelago*, and had become well-known. Frederick Boyle was well read and informed, so we might ask why he hadn't mentioned the Wallace connection in his later work. Perhaps it didn't fit his literary flow? Or maybe he wanted to maintain the reader's focus on his own exploits?

My conclusion: The timing of his engagement, his competence, and fashion sense suggest Ali Kasut was the Ali who accompanied Wallace.

If so, a larger question beckons: Where did Ali go after finishing his employment with the Boyle brothers? Did he stay in Sarawak? Or did he return to Ternate?

Like all of us, Ali faced a number of crossroads — should he have turned left, should he have turned right, should he have sat still and let the universe decide? Perhaps his thoughts went something like this:

After the Boyle brothers left, Ali was once again alone, unemployed, and a bit of a lost soul.

Life is full of decisions, Ali realized. You are presented with choices, some better than others. Each choice has implications; you can't ask for a do-over. Ali wasn't a philosopher, and he had obviously never heard of psychoanalysis, but both disciplines were background noises in his active mind. And all the options he had were overshadowed by one major persona: his wife.

He had gotten married in Ternate, an action that surprised him but was forced upon the young man after getting a young lady named Fawziah pregnant. It had been the first time for both of them, a quick forgettable fumble in the woods surrounding Ternate's crater lake. But family reputation was important, and Fawziah's father got out the proverbial shotgun. The couple lived in a small house. They quickly had two girls. They grew to tolerate each other.

Those were, perhaps, the facts behind Ali's self-questioning. He knew what he should *do — return to Ternate and look after his wife and children.*

Romantic love didn't have take up much real estate in Ali's mind, but if he was honest with himself, he realized he actually didn't care that much for Fawziah. She was fussy and petty, talked too much, and was always bugging him for money to buy new sarongs. She had little interest in the exotic foreign adventures Ali had experienced. She was a lousy cook, and she was getting fat. Fawziah was so different from the independent and spirited girls of Sarawak with their feathers and dances and bird omens, and she was the polar opposite to the demure, elegant, and meticulous girls of Java. He had never actually been with a girl from Sarawak or Java, of course, but he had the vigorous imagination of a teenager.

And in Sarawak he was rich. He had the boat the Boyle brothers had given him. He knew his way around, he had connections, he could become a rich trader and have his pick of women.

But his little girls back in Ternate were adorable. The responsibility to be there for them weighed heavily on his heart.

And Tuan Wallace had told him often about a man's responsibility to others. True enough, but Tuan Wallace hadn't taken responsibility for Ali at the very end, only throwing some gifts his way as he waved goodbye.

Ali stood on the dock at Kuching and watched the boat carrying the Boyle brothers to Singapore sail off to the west. He had money in his pocket. He had a fine boat. He had employable skills. And he was the loneliest man in Borneo.

CHAPTER 13

Ali Returned to Ternate?

There are several convincing morsels of evidence that Ali settled in Ternate.

Wallace used the town as his eastern Indonesia base camp for some three and a half years, where he recovered from voyages to distant islands and prepared for his next adventure. He wrote fondly about the creature comforts he enjoyed in Ternate:

> In this house I spent many happy days. Returning to it after a three or four months' absence in some uncivilized region, I enjoyed the unwonted luxuries of milk and fresh bread, and regular supplies of fish and eggs, meat and vegetables, which were often sorely needed to restore my health and energy. I had ample space and convenience for unpacking, sorting, and arranging my treasures, and I had delightful walks in the suburbs of the town, or up the lower slopes of the mountain, when I desired a little exercise, or had time for collecting.[141]

He particularly appreciated the availability of clean water:

> A deep well supplied me with pure cold water – a great luxury in this climate.[142]

Wallace twice refers to Ali getting married and establishing a family in Ternate.

Wallace first refers to Ali's marriage in a letter to Samuel Stevens from Ceram on November 26, 1859:

> [My] best [man, referring to Ali] is married in Ternate, and his wife *would not let him go* [with Wallace to Ceram]; he, however, remains working for me and is going again to the eastern part of Gilolo [italics Wallace].[143]

Many years later, in his autobiography, Wallace again refers to Ali's marriage and suggests that Ali's wife had loosened her grip on her husband's travels:

> During our residence at Ternate he married [probably in early 1859], but his wife lived with her family, and it made no difference in his accompanying me whenever I went till we reached Singapore on my way home.[144] [145]

Trustees of the Natural History Museum, London. Photographic image manipulation by Kelly Cleary.

A computer-generated image of how Ali might have looked at approximately age 67, when Thomas Barbour met him in Ternate.

Earl of Cranbrook and Adrian G. Marshall agree that Ali returned to Ternate, stating unequivocally "After ARW [Alfred Russel Wallace] departed from Singapore, Ali returned to Ternate to rejoin his wife and, perhaps, a young family."[146]

The closest we have to a smoking gun confirming that Ali settled in Ternate came from American naturalist Thomas Barbour, who, in 1907, visited Ternate and claimed he met Ali. We know that Ali was about 15 when he met Wallace in 1855, so he would have been around 67 when he met Barbour, a relatively old man but well within the limits of possibility.

Barbour mentions meeting Ali three times.

In 1912, Barbour, director of the Museum of Comparative Zoology at Harvard University, wrote in a scientific paper:

> I showed a Ceram specimen of *L. muelleri* [*Lerista muelleri*, common name: wood mulch-slider][147] to many intelligent natives of Ternate, including indeed Ali, the faithful companion of Wallace during his many journeys, now an old man, and all agreed that they had not seen such a lizard before.[148]

Brian Bush

The wood mulch-slider, *Lerista muelleri*, that Barbour said he discussed with Ali in Ternate, Indonesia, isn't found in Indonesia. The entire genus Lerista is now thought to exist only in Australia, so it is unclear whether Barbour was (incorrectly) talking about this specific animal or discussing (and mis-identifying) a different species.

In 1921, in another scientific paper, Barbour wrote:

> On the day of my walk to the Ternate lake an old Malay spoke to me; he had long forgotten his English, but he tapped his chest, drew himself up and told me he was Ali Wallace. No lover of "The Malay Archipelago" but remembers Ali who was Wallace's young companion on many a hazardous journey. After my return a letter from Mr. Wallace speaks of his envy of my having so recently met his old associate.[149]

And in his autobiography of 1943, *Naturalist at Large*, Barbour wrote the most detailed account:[150]

> Here came a real thrill, for I was stopped in the street [in Ternate] one day as my wife and I were preparing to climb up to the Crater Lake. With us were Ah Woo with his butterfly net, Indit and Bandoung, our well-trained Javanese collectors, with shotguns, cloth bags, and a vasculum for carrying the birds. We were stopped by a wizened old Malay man. I can see him now, with a faded blue fez on his head. He said, "I am Ali Wallace."[151] I knew at once that there stood before me Wallace's faithful companion of many years, the boy who not only helped him collect but nursed him when he was sick. We took his photograph and sent it to Wallace when we got home. He wrote me a delightful letter acknowledging it and reminiscing over the time when Ali had saved his life, nursing him through a terrific attack of malaria. This letter I have managed to lose, to my eternal chagrin.[152] [153] [154] [155]

This should be sufficient evidence, but there are questions about Barbour's claim.

Barbour wrote that "We took [Ali's] photograph and sent it to Wallace when we got home." However, it seems from Wallace's reply, below, that Barbour was mistaken — he had either *not* taken a photo of Ali or had sent Wallace a different photo:

> Thanks for sending me news of... my "boy" Ali – a photograph of whom would have been more interesting to me than those of the men of Dorey, who are pretty nearly as I left them 50 years ago.

The Ernst Mayr Library and Archives of the Museum of Comparative Zoology, Harvard University

Thomas Barbour (1884–1946), a well-travelled and respected American herpetologist who was director of the Harvard Museum of Comparative Zoology, provided the most convincing evidence that Ali spent his "retirement" years in Ternate. Barbour wrote of meeting a "wizened old Malay man" who was versed in natural history and identified himself as Ali Wallace.

Barbour seems to be disassembling. The only photo in existence we have of Ali is the famous one of him dressed in European clothes, taken in 1862 in Singapore. The purported photo of Ali that Barbour claimed to have taken and intended to send to Wallace has never been found in either Wallace's or Barbour's archives. Perhaps Barbour had *wanted* to take a photo of Ali. Perhaps he *had taken* a photo, but it didn't come out. Perhaps he *thought* he had sent a copy to Wallace but hadn't. We'll never know. Either way, a librarian at the Harvard University archives, where Barbour's papers are housed, terms Barbour's claim as an example of him "misremembering," which, if true, might put Barbour's entire reminiscence in doubt.

So, one might speculate that instead of meeting Ali, Barbour actually met one of the Ternate-based collectors who travelled with Wallace who said "I *knew* Ali and Wallace." Perhaps Wallace's very polite statement that he would have preferred a photo of Ali indicates that Wallace desired confirmation of Ali's existence in Ternate. Wallace doesn't ask Barbour for any news about Ali or how to contact him, which might indicate he was skeptical about Barbour's claim. Wallace did not include the Barbour information about meeting Ali in the New Revised Edition of his autobiography that came out in the Autumn 1908 (the Wallace to Barbour letter is dated February 21, 1908). Clearly there had been time to include the Barbour information in a footnote. Or in the reprint of *The Malay Archipelago* in 1913.

Another piece of informed speculation is this: We know Ali was a skilled collector and taxidermist and that Maarten Dirk van Renesse van Duivenbode, Wallace's rich and powerful friend in Ternate, ran a successful business selling birds-of-paradise to European collectors. It is not impossible that Wallace told van Duivenbode about Ali's skills, and after Wallace left, Ali easily found work with the Dutch bird trader.

"With Ali's skills, it is logical that he would easily have found employment with van Duivenbode or another European trader" according to Dutch historian Marc Argeloo, who has studied the 19th-century Ternate bird trade. "We haven't yet found written evidence that Ali worked for van Duivenbode, but it seems plausible."

Conclusion? As with so many "facts" about Ali's life, there is little certainty about where Ali ultimately settled. But my feeling is that after he left the Boyle brothers in Sarawak he returned to Ternate.

This admittedly arguable conclusion is based on three pieces of information and circumstantial evidence. Wallace wrote several times that Ali had a family in Ternate. Thomas Barbour, a well-known naturalist and academic, wrote three times about meeting "Ali Wallace" in Ternate. And Ali's skill in bird collecting and taxidermy would have enabled him to easily find a job with the European bird traders in Ternate.

IMAGINED CONVERSATION: "I HOPE THE CAMERA WORKS"

Thomas Barbour, his wife, and several assistants walk along a dirt road just outside the town of Ternate, eastern Indonesia. They are headed to the Crater Lake, in the center of the island, on a collecting mission. Barbour is a multi-faceted naturalist, but herpetology is his first love — in particular, he takes an inordinate interest in snakes and amphibians — slithering creatures that hold little interest for most people.

A short older man runs after them. He wears a clean, but faded sarong, a cotton shirt that once was white, and a songkok, a Malay version of the Middle Eastern fez. He is out of breath, having jogged up the hill. He hadn't wanted to call after them — that wouldn't have been polite — but he does want to speak with this white man.

"Excuse me, sir," Ali says in English, still out of breath.

"Selamat pagi," Barbour says in Malay.

"Ah, you speak Malay," Ali says in the same language. He switches back to English. "You are collecting?"

Barbour and Ali have a conversation, assisted by Barbour's Indonesian assistants where necessary. They discuss the best places to find lizards and frogs.

"You are very knowledgeable," Barbour says.

Ali beams.

"And may I know your name?"

"I am Ali Wallace," Ali says, with more than a hint of pride in his voice.

Barbour and his wife look at each other. Couldn't be. But yes, here he is.

"You are Ali, who accompanied Alfred Russel Wallace?"

"You know my Tuan, sir?"

"I've never met him. But everyone knows Wallace. He is a great man."

"Yes. And how is he? You are from the same kampong?"

"Er, no. I live in America. Wallace lives in England."

Ali rolls the two words in his mouth like hard candies, savoring the taste of the names. America. England. "Does he ask about me?"

"I'm sure he thinks about you, Pak Ali. You were very important for him."

Ali, to the surprise of both men, begins to cry. "I am an old man, Tuan. Thanks to God, I have lived so long. I live, every day, praying to Allah, hoping for a letter from Tuan Wallace . . . "

Barbour's wife, Rosamond, also starts to cry. "We'll be sure to tell him we saw you." She doesn't ask if Ali has a mailing address.

Ali continues, in Malay, as if he hasn't heard her. "Tuan Wallace left; I came back here. He could have sent me a letter care of the Dutch administrator. They have a good mail service." He is rambling now. "Tuan Wallace wrote so many letters, so many, to his friends in England. Does he have a family? Does he speak about me? Is he in good health?"

"Let's take a photo, Pak Ali," Barbour says. "I'll send it to Wallace and tell him we met." Then, as he was setting up the equipment, muttered: "My camera's been acting up; I hope it turns out . . ."

CHAPTER 14

The Search for Ali's Descendants

I'm satisfied that Barbour's encounter and Wallace's comments about All having a family in Ternate make it likely that Ali settled in Ternate after he and Wallace parted ways in Singapore.

Perhaps not enough evidence for a court of law, but certainly enough for a quest.

My logic:

After leaving Wallace in Singapore, Ali probably returned to Sarawak for a while in the guise of Ali Kasut (Chapter 12, "Ali Returned to Sarawak"), then continued on to Ternate as an important and relatively wealthy man. He had travelled widely in the employ of a tall, respected, undeniably quirky Englishman. He had his photo taken wearing European clothes, and he was a raconteur, which signals he would have been a "personality" widely known in his community. He had tall tales to tell and wasn't shy about recounting them to anyone who would listen. Wallace recalled that while in Singapore Ali had:

> seen a live tiger [and] made much of his knowledge when we reached the Moluccas, where such animals are totally unknown. I used to overhear him of an evening, recounting strange adventures with tigers, which he said had happened to himself. He declared that these tigers were men who had been great magicians and who

> changed themselves into tigers to eat their enemies ... These tales were accepted as literal facts by his hearers, and listened to with breathless attention and awe.[156]

Wallace wrote that Ali had a family in Ternate. Surely, I thought, someone would have a memory of great-great-granduncle Ali.

I went to Ternate island and tried to encourage the mayor, journalists, and academics to find Ali's descendants.

I held a few press conferences, but at the end of the day there was no progress. I was frustrated that my seemingly Quixotic quest seemed doomed to be defeated by the black hole of enthusiastic Indonesian inertia.

Perhaps it was personal. I thought they didn't like me, or they weren't interested on expending effort on someone else's idea.

Paul Spencer Sochaczewski

Paul Spencer Sochaczewski

Perhaps my lack of success could be that at the time of my visits Wallace's (and by association, Ali's) star was shining less brightly, at least according to the judgment of city officials. The medium-sized street where his house was ostensibly located, which had in the early 21st century been renamed Jalan Alfred Russel Wallace (Alfred Russel Wallace Street), had been reverted to its original name, while a narrow passageway off the street near the alleged house was given the name Lorong A. Wallace (Alfred Russel Wallace Small Alley).

I suggested the officials could place an article in the local paper; I even volunteered to write it. "Let's show Ali's photo and story to the village elders," I suggested. "Let's get a university student to write a thesis on the search for Ali." I pointed out that this search could generate national, even international news, and be a stimulus for the tourist business and a catalyst for conservation. It would certainly help build local pride, which is always a good thing for elected officials to bask in. *Bagus*, everyone said. Good idea. Lots of enthusiasm, no action. Indonesian apathy? A dearth of intellectual curiosity? A reluctance to pursue a "foreign" idea?

I gave lectures to history and biology students at universities in Ternate and neighboring cities and encouraged them to seek Ali. A few students I spoke with privately expressed polite interest, but quickly zoned out when I explained that for their research they would have to speak with old folks in the villages and go through dusty records. "But that sounds like ... field work!" they would explain, before politely excusing themselves.

This quest, like many others, remains a work in progress.

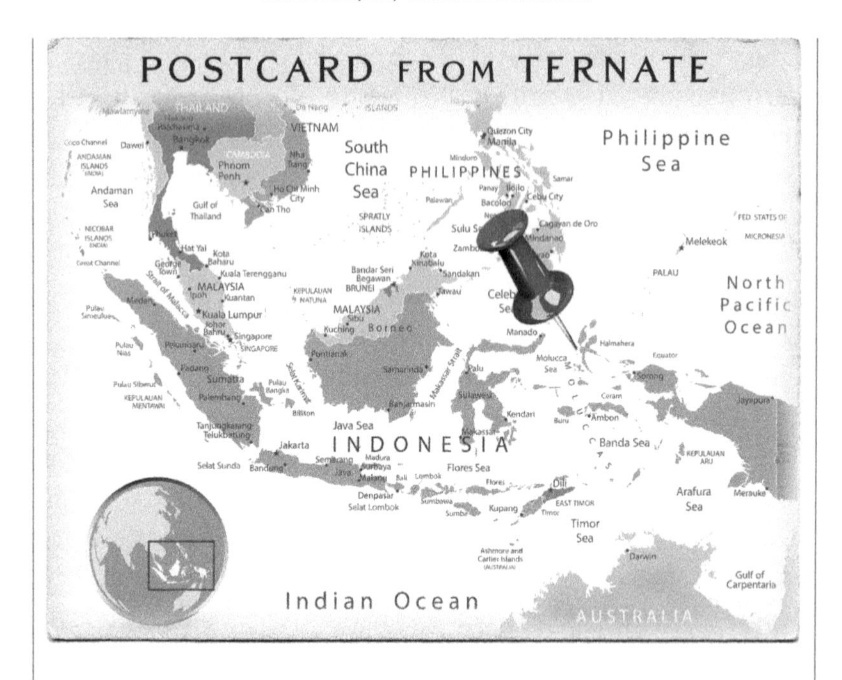

It's hard to overstate the remarkable influence two isolated east Indonesian islands have had on history.

Banda Neira was the source of nutmeg. Ternate was the source of cloves. The search for the origin, and eventual control, of these spices, and eventual control, spurred the great European expeditions and colonial activities of Portugal, Spain, Britain, and the Netherlands.

Although tiny, these islands are postcard-worthy jewels, with a fuming volcano smack in the middle of each island.

Wallace lived in Ternate, on and off, for some three years. It was his base camp, his refuge, his stability in a life of challenges. And Ternate is likely where Ali "retired" after seeing off Wallace in Singapore in 1862.

The Manhattan-sized island of Ternate includes a small city of 200,000 people. I visit every year or two. Years ago I climbed Gamalama volcano, a dramatic, often-sputtering peak that is sacred for the royal family of Ternate. Now I go to see friends, eat seafood, soak up the atmosphere of Wallace's residence, and try (without much success) to drum up interest among local historians and students to look for Ali's descendants.

More recently, I was part of a group of researchers who were trying to locate the precise location of Wallace's house in Ternate. We succeeded by following the indications he wrote in *The Malay Archipelago*, then searching for the key clue — a deep well with clean, fresh water.[157]

Paul Spencer Sochaczewski

Graffiti of Wallace (sometimes paired with images of Ali) can be seen in the neighborhood where Wallace's house was located.

Antiqua Print Gallery, Alamy Stock Photo. From Histoire Générale des Voyages, ou nouvelle collection de toutes les relations de voyages par mer et par terre.

A 1761 French copperplate print of Ternate featuring Gamalama volcano and a busy harbor.

SECTION V

Trying to Speak with the Spirits of Wallace and Ali

CHAPTER 15

Might the Ghosts of Wallace and Ali Help Me Unravel the Unanswered Questions About Ali's Life?

"Why don't I just go to the source?" I asked.

I was having a seafood dinner with Ofa Firman, a son of the late sultan of Ternate, and was venting my frustrations about trying to find the descendants of Ali.

"And by 'source' you mean what?" Firman asked as he passed me a plate of tamarind prawns.

I suggested that together we create a new form of historical research. Rather than examining academic tomes, sifting through dusty Dutch records, and going door-to-door in the Malay settlements, why not talk to Ali himself? "That sounds like a fine idea," Firman said, "but there's a slight hitch. Ali is dead. Which means he's a ghost." A *hantu* in Indonesian terms.

"Ternate is full of mediums," I said, passing him a platter of chilli crabs, a specialty of Royal's Resto and Function Hall. "Let's talk to Ali himself and get the true story."

Let's call it History by Hantu. If my History by Hantu works, what might we learn? Will we unravel the "mystery of Ali?" (On a broader scale, could such investigations answer the great historical mysteries? What if we could speak with Lee Harvey Oswald and ask if he acted alone, and for what reason? Or get into the head of Napoleon, or Cleopatra, or Genghis Khan? Wouldn't scholars love to ask Shakespeare if he really wrote all those plays and sonnets? Or ask Mary, mother of Jesus, about that virgin birth story?[158] [159] [160] [161]

"I know just the guy," Firman said.

I clearly needed spiritual intervention. Not the religious kind but the purported wisdom that comes from speaking with dead people.

Do you believe in spirits?

I don't. I don't believe that we have an eternal soul, spirit, energy, persona that lives on after we die.

THE

CENTRAL ASSOCIATION OF SPIRITUALISTS

(With which is incorporated the British National Association of Spiritualists, Established 1873).

:o:

This is to Certify

that A. R. Wallace Esq., F.R.G.S.

has been elected an Honorary Member of this Association.

Reg. No. 39 ... *President.*

Date 13th June, 1882. ... *Secretary.*

Trustees of Natural History Museum, London

Wallace's 1882 certificate designating him an honorary member of the (UK) Central Association of Spiritualists.

Alfred Russel Wallace, however, was an ardent spiritualist, and he actively participated in seances and psychic demonstrations that were en vogue in Victorian England. He attended such events whenever he had the chance, accepting that the disembodied bodies, independently moving ouija boards, mysterious bell ringing, and fuzzy photographs of his dead mother hovering next to him, were real.

He was a careful, deliberate, well-respected scientist. He was outspoken when he chose to make a statement (but reticent when it came to challenging Darwin's priority in developing the Theory of Evolution by Natural Selection). But he was gullible when it came to explaining the inexplicable psychic phenomenon he witnessed, in spite of the evidence (convincing for most people) that he was being tricked by charlatans.

He famously wrote:

> The so-called dead are still alive. Our friends are still with us. They guide and strengthen us when owing to absence of proper conditions they cannot make their presence known.[162]

In an 1887 lecture, Wallace concludes:

> It further demonstrates, by direct evidence as conclusive as the nature of the case admits, that the so-called dead are still alive — that our friends are often with us, though unseen, and give direct proof of a future life, which so many crave, but for want of which so many live and die in anxious doubt.[163]

Wallace believed there was sufficient evidence to justify his belief:

> During the last sixty years evidence has been accumulating in every part of the world which affords demonstration that the so-called dead have never really died at all, but have passed into a new and higher stage of existence. Many of these are able to communicate with us and most of them assure us that when they wake from the sleep we call death they find themselves much more alive than ever they were before. And this is only what we might expect; for

> we all feel that our mental faculties are to some extent clogged and stifled by the garment of flesh, and that only when in the most perfect health do our higher faculties attain their fullest expression.[164]

But he noted that some things cannot be explained by science:

> We have all kinds of phenomena which are inexplicable even to the scientific mind, except on a spiritualistic hypothesis.[165]

College of Psychic Studies, London

A photograph of Alfred Russel Wallace with a spirit which he described as "a male figure with a short sword,"[166] taken by a charlatan photographer, Frederick Hudson.[167] Wallace firmly believed in communication with spirits, and it is highly likely that Ali did as well, although their cultural filters no doubt perceived the phenomena differently. This belief continues in our technologically advanced world — one out of every five Americans believes that the deceased can communicate with those who are still alive.

Today, Wallace's views resonate with large numbers of people.

Spiritualism is alive and well. A recent poll found that 65 percent of Americans believe that after death people go to heaven, hell, or purgatory, seven percent believe they go to another dimension, 6 percent believe they are reborn on Earth, and two percent believe they become ghosts. Strikingly, one out of every five Americans believes that the deceased can communicate with those who are still alive.[168 169]

This certainty in something that can never be proved requires an individual to buy into the following sequence of logical assumptions for illogical happenings. I call this the Six Basic Tenets of Spiritualism:

1. Each of us has an unseen, ethereal, eternal entity called a soul. Or spirit. Or energy.
2. This soul/spirit/energy lives on after the physical body dies. (The location is unclear. "Everywhere and anywhere" seems to be the best assumption.)
3. This soul/spirit/energy is generally benign. Except in horror films.
4. This soul/spirit/energy is willing to contact still-living descendants. Such contact is rarely in the form of a complete or coherent discussion.
5. Each of us theoretically has the power to contact this soul/spirit/energy. But just as most of us wouldn't attempt to fix a plumbing problem in the house, it's generally more effective to work via a skilled medium/channeler/psychic/shaman who can act as a go-between.
6. There is no guarantee that we can contact a specific soul/spirit/energy at any given time.

What happens when we die?

Is it the end? Or the beginning of the next phase?

It's a binary question: Annihilation or metamorphosis?

At its essence, belief in spirits is more than simply trying to say hello to ancestors who have "passed." It's also a confirmation of our own complex presence, a basic validation that some essence of ourselves will survive physical death. *I have a value and existence beyond my body.* This belief enhances our desire to leave a legacy. How are we otherwise remembered by later generations for whom our lives become ever-distant folklore?

Most mediums I told about my desire to speak with "a specific historical character" scoffed at my presumption. *Spirits will only talk to you if they have a connection, and if it suits them to make an appearance. You can't dial up a spirit like a LinkedIn connection.*

I could interpret this caveat in several ways.

I failed several times to contact the spirit of my father, Samuel. We had (or have, it's odd trying to figure out the correct tense) a strong emotional attachment. But what if he was busy; perhaps I was trying to reach him on a Tuesday night when he was playing poker?

Or perhaps, if there was no contact, his spirit might have sailed off into another dimension and broken all connections with this mundane planet.

Or he might have been reincarnated as a bumblebee, a bonobo, or a bonze.

But there is another explanation why my father "wasn't home" when I called. It's due to the possibility that the idea of speaking with the spirits of dead people is a hoax, and mediums don't want to promise something they can't deliver. I have been to numerous public sessions. The medium does not ask anyone in the audience "is there anyone in particular you want to speak with?" No, the medium closes his or her eyes, then says: "I'm getting a white dress. Does that mean anything to someone?" And a person in the audience might say, perhaps timidly but with a tinge of excitement, "My mother liked to wear a white dress," to which the medium would say, "And she died recently?" And they're off in a dialogue in which the medium controls the joystick. If the medium "feels" roses in a garden, then someone might say, "My father loved watching Gardeners' World." But sometimes the medium might offer a clue from the afterworld ("I'm seeing lots of books") that doesn't generate a positive response. The medium is likely to explain away the lack of audience response by announcing, "There are several spirits who want to talk, so many messages, they're talking over each other, the messages are jumbled."

The endgame is always the same. *Your mother loves you. She wants you to let go and get on with your life. She will look after you. You'll be fine.* People seek mediums to deal with grief, to have the reassurance that a loved one is somehow, still "alive," albeit in a different dimension. Reassurance and comfort are the essential tools of the skilled medium.

(For more on my experiences with mediums trying to speak with Wallace, Ali, my father, a female man-hating ghost in Borneo, nature spirits in Burma, and the mystical mermaid who is the consort of the sultans of Java, see *Dead, But Still Kicking*.)

Why does spiritualism have such traction?

The euphemisms "passed away," or "crossed over" have the "just maybe, fingers crossed, I hope-I hope" optimistic ambiguity of the mother of a 16th-century sailor setting off around the world. But the harsh term "death" emits the irreversible finality of a tympani roll followed by the clash of the cymbals, the end-times "whoosh-thwack-thump" of a guillotine. A belief in spirits is the antithesis of ashes to ashes, dust to dust. A belief in spirits is the anticipation (often fueled by a religion that promises a form of "heaven") that life continues after death, preferably in a better, safer, kinder, less-stressful place.

I thought it might be fun to give it a go. Why not try to communicate with the spirits of Wallace and Ali? Surely if anyone would, Wallace would be receptive to help me fill in the blanks.

The results were inconclusive, as I expected. Sometimes (okay, often), my encounters with mediums were as silly and entertaining as a Marx Brothers movie. But on occasion they were also disturbingly insightful. And it's those infrequent disturbing occurrences, which appeared as dramatically as a flash of lightning, that gave me pause. I'm still a non-believer, but in the dark recesses of my mind I remember bursts of illumination that forced me to think: *How does she know that? What the hell is going on here?*

I am an Agnostic Spiritualist (but also identify with Charles Dickens, who described himself as a "fascinated skeptic"). I appreciate science. I don't believe in organized religion. I think that when we die, that's it. Life is a beer commercial — live life for all its gusto because you'll only go around one time. Carpe diem and all that.

And yet.

I am convinced that, to paraphrase Hamlet, there are things in our world that cannot be explained using Cartesian logic.

Most of the readers of this book will have had a Western education and upbringing that stresses evidence-based science. Yet when you ask well-educated people, "Do you believe in spirits?" they might prevaricate. "Not really, but I'm not sure. There are things that happen that we can't explain."

As with our personal "story," a belief in spirits is enhanced by what psychologists call a self-perpetuating loop. We invest energy in a belief, and through repetition and endorsement by others, the idea increases in value; this is termed embedment: *The more we repeat some idea, the stronger it becomes, and the harder for us to eliminate. We have invested energy in our "story," and that story becomes increasingly real in our minds.* The malleable clay of an idea has hardened into an unshakeable belief. Put another way, author Neil Gaiman said: "Things need not have happened to be true," an observation as relevant to politics as it is to the spirit world. We get entangled in wishful thinking — we *believe* something is true because we *want* it to be true.

The belief in an existence after death is the great promotional offer of religious proselytizers to ensure adherence to a set of beliefs; it's the most important element of religion marketing. The majority of people in the world profess some kind of formal religion. But what is a religion but a set of rituals and rules that uses miracles and parables to accomplish two things? First, most religions promulgate the existence of a one or more Super-Entities usually either a form of Hairy Thunderer or a Cosmic Muffin, which has oversight and free reign over everything. And second, to ensure fealty to this Super-Entity, religious leaders say that each of us has a soul of some kind. This soul, the priests suggest, can survive the death of the body and, provided we collect enough karma points, permit us to enter a form of happy afterlife. The dilemma is that, unlike frequent flyer credits (we know when we have enough points for a flight to London), we never know how many karma points we need to enter heaven. But the downside is that without the requisite credits, we might descend into the eternal furnaces of hell or be forced to try again. And again. And again, until we get it right. Salvation or damnation. It's a binary option.

Regardless of whether we are religious (and the idea of eternal souls is an equal opportunity concept embraced also by countless atheists, agnostics, laborers, teachers, politicians, Pagans, Animists, free-thinkers, pendulum-swingers, tarot enthusiasts, astrologers, and people who commune with fairies, many of us believe, at least partially, that ghosts, spirits, and djinns exist.

Is belief in a spiritual existence insight or delusion? Our ability to reason, our all-powerful left brain that has been nurtured by a diet of algebra and the red pencil of stern Mrs. Olsen in fifth grade who scolded us when we misspelled a word, says there's no such thing as spirit. It's an illusion.

You can't prove it, therefore it's myth.

Yet plenty of people, often intelligent, well-educated people, believe in spirits, just as they believe in love, intuition, and superstition.

Let's put it another way. An acceptance of the idea of spirits and mediums involves "letting go" of our materialistic worldview and admitting there are some elements of our existence that we don't understand and can't prove. It's like quantum mechanics, which arose from theories to ex-plain observations that could not be reconciled with classical physics. Quantum mechanics suggests that consciousness (not matter) is the primary mover and creator of reality. You might be frightened and confused and want such stuff to remain in the genie's bottle. Or you might be thrilled by the unprovable idiosyncrasies of the universe and be willing to explore your psychological horizons.

Most mediums told me that they can't promise to dial-in with a specific dead person. But while researching *Dead, But Still Kicking*, I met with ten mediums who had, on my request and in my presence, "conversations" with Alfred Russel Wallace.

The results were sometimes ridiculous, often vague, but occasionally startling. In some cases they were so precise (judge for yourself in the following transcripts) that it's hard not to believe the medium did a pre-séance Google search to learn more about both Wallace and me. I don't believe in this talking-to-ghosts stuff. But how can I explain the psychic's unsettling drawing of Wallace and her goose-bump response to my mention of the name "Darwin." Am I being conned, or is there something going on beyond my comprehension?

I offer these evocative conversations (one allegedly with the spirit of Wallace and six allegedly with the spirit of Ali) as intriguing diversions, not as historical evidence of anything greater than my own curiosity.

Make of them what you will.

My conclusions?

Degree of credibility: somewhere down by zero.

However, if I suppress my natural cynicism, I can imagine that I have a strong relationship with Wallace – some friends suggest he's been my mentor, a wise uncle-type who gives me a nudge, a pat on the back, and a hard kick when I need it.

But it seems I do not have a good psychic connection with Ali. Perhaps that is why he remains in the shadows, a mostly hidden, largely reticent, illusionary spirit.

On the following pages I offer one of my ten channeling experiments with Alfred Russel Wallace and six experiences trying to speak with Ali. (For full transcripts and commentary, see *Dead, But Still Kicking*).[170]

CHAPTER 16

Good News! Alfred Russel Wallace Is My Spirit Guide, My Mentor, My Pal

I find June-Elleni Laine through a mutual friend at the College of Psychic Studies in London.

I am careful to tell her only that I want to speak with a 19th-century historical figure. I do not give her names or other information.

Before we start, she tells me about her worldwide clients, and explains that there are no guarantees. She says she draws portraits of dead folks and enjoys ongoing conversations with dead famous people, such as Leonardo da Vinci. The skeptic in me thinks that she might glean clues about my interests and intentions from this type of conversation. And I'm visible via a Google search if someone wanted to really delve into my background (and locate my interest in Wallace).

> Like Brigitte Favre (a Geneva-based psychic who also channeled Wallace for me), June-Elleni Laine is a stickler for honesty and transparency. *"I'm a skeptic myself. I can't do any work that can't be validated."*

I ask her permission to record the session and use excerpts in my book. She agrees and adds that she has been misquoted in previous published pieces and asks me to be accurate.

We agree that I will pay her fee by PayPal and speak again in a few days for a full-fledged session.

And then, just before we break the Skype call, out of nowhere, June-Elleni says:

> *"This gentleman you want to speak with. He's your spiritual guide."*

I tell her that I thought a spiritual guide is the intermediary between the sitter and the spirit. Sort of like a celebrity agent, the one who makes the contact with the spirit, who ensures that both parties want to chat, who protects the sitter if the vibes aren't positive.

> *"No, that's a gatekeeper. This gentleman is your spiritual guide."*

Other mediums noted I have a connection with Wallace, and vice versa, and that Wallace is like a "soul brother" (or Mentor to use Joseph Campbell's terminology) who will guide me through life. But those mediums stressed that this assistance is only available if I ask.

> *"I'm picking up a man who is a humanist. I think you told me that,"* June-Elleni says.

"No, I certainly didn't tell you that," I say. But this reinforces my feeling that mediums employ a fair degree of telepathy or intuition. Perhaps June-Elleni senses, after an earlier conversation, that I too might be a humanist, and she guesses I would want to speak with a like-minded individual. "But yes, you could call him a humanist."

> *"He wants to come through to you. Together you can make a difference."*

Sounds like another joint venture opportunity, similar to an earlier spirit conversation I had in Indonesia when Moses gave my friend Yan Mokoginta and me instructions to negotiate peace in the Middle East.

> *"Great sense of moral obligation."*

That sounds like Wallace.

> *"I recall reading something — the way human beings torture themselves is funny as hell. No, it's funnier.' Would your friend be amused by that?"*

I sense she is fishing, but I have to agree. "Absolutely."

> *"He dressed like a gentleman. Some kind of boarding school attire."*

Now this is interesting. I don't think Wallace was too concerned about how he appeared (for one thing he rarely had enough money to buy elegant clothes), but he was very concerned with being *improperly* dressed. In his autobiography he recalls how embarrassed he was when, as a gawky, rapidly growing youngster he used his sleeve cuff to clean his school slate. This so irritated his mother that she made calico over-sleeves. He refused to wear them, until his mother asked the teacher to order him to do so. This shame, in front of the entire class, was "perhaps the severest punishment I ever endured," he wrote. Wallace specifically used the term "loss of face."

We agree a date for the next session. But I'm impressed that our conversation with Wallace has already begun.

Left: Wallace with what he believed was the spirit of his dead mother. Photograph by Frederick Hudson.
Right: Alfred Russel Wallace's mother, Mary Ann Wallace (née Greenell).
Marchant, James (1916). LLP collection/Alamy Stock Photo.

Alfred Russel Wallace learned mesmerism and phrenology as a teenager. After his return to England following eight years in Southeast Asia, he became a proponent of spiritualism and attended numerous seances. A noted photographer of ghosts, Frederick Hudson took this photo in 1874 purporting to show Wallace with his dead mother. Wallace was convinced it was a genuine spirit photograph, declaring: "I see no escape from the conclusion that some spiritual being, acquainted with my mother's various aspects during life, produced these recognisable impressions on the plate."[171 172] Hudson was later discredited for manipulating images in the darkroom, even posing as a spirit to create some of his fakes.

June-Elleni looks relaxed, sitting in a simple room, family photos behind her. She has a long, attractive face, wears a simple grey and white cable-knit sweater, dark curly hair loosely tied behind, and a minimum of jewelry — but what she does wear is simple, refined and elegant, what appears to be gold and diamond stud earrings, thin necklace, and a discrete ring.

> *"Just so you're clear. I'm not sure who will appear. Do you want to give me a name to try to zero in?"*

Up to this point I have not given June-Elleni any clues except my desire to speak with "an historical figure." I don't want to give away Wallace's name, in case she has knowledge about him. I simply say "Alfred."

"Now, this is interesting. Mothballs?"

"No idea."

"As before, I see him in a tweed jacket. One of my favorite teachers wore a tweed jacket, Mister Elliott. So possibly Alfred was a teacher?"

Well, Wallace taught when he was younger, for a brief period, but more important he was a teacher in the larger sense of opening people's minds and forcing them to think differently about themselves and their world. But, of course, that description could apply to almost anyone. I consider this a square-peg-in-a-round-hole moment — if you pound hard enough, you can make almost any psychic pronouncement fit your worldview.

"He's been trying to contact you. I get the number fifteen. Could be fifteen years or months?"

I suppose I could make either number work if I twisted and strained and discombobulated, but nothing jumps out at me.

"Did he live in the 1800s?"

I try to remember if I had described him as a 19th-century personage. I think I had. Damn. "Yes."

"He was a pioneer, writing about new ideas that were later validated."

"Absolutely."

"I am hearing the word 'whales.'"

"The animal? That doesn't ring a bell."

June-Elleni corrects: *"Could be the animal as a metaphor for something large and almost mythical. Or it could be the sound of the word, maybe the nation Wales."*

Ah, the power of homonyms. Wallace was born in Usk, a small village in Wales, just across the border from England.

"Good singing or speaking voice." She admits the Wales-connection got her thinking of the singer Tom Jones.

The first of numerous mind-wandering statements. A conscious diversion?

June-Elleni is busy drawing. She continues to draw during our hour-long conversation. *"Ooh! Quick spurt of energy just now. I'm getting a hairy face."*

"Could be." I don't want to give too much away. I think, but don't mention, that Wallace had a luxuriant beard for most of his adult life.

"Not a big man in stature, average or below-average height."

Here June-Elleni makes her first big error. Wallace was tall, six-foot one-inch.

Then she self-corrects and reboots. She adjusts course, like a sailor changing tack, a common tactic of mediums. June-Elleni stresses, as most mediums do, that sometimes the information she receives is not to be taken literally but should be viewed as metaphors. *"Were there times when people disagreed with him or made him feel small?"*

I reluctantly admit her self-correction makes sense. "Yes."

"And this resulted in anger, disappointment. No, more frustration than anger." "He couldn't understand the mentality of people who wanted to bicker just for the sake of arguing." She returns to the hairy face. *"Are you aware of a beard?"*

Is she reading my mind? That, by itself, would be a remarkable achievement. "Yes."

"I smell smoke. Did he smoke? Or was he around people who smoke?"

No idea, but I don't think he smoked. Could this refer to the smoke arising from the brig *Helen* when it caught fire on the return voyage to England from Brazil? If I want to stretch things, I could imagine an open cooking fire while he was living rough in Asia.

She's back to the hair. "*Thin on top?*"

"Not especially. He died with a full head of hair." Of course, Wallace might have *thought* his hair was thinning. How literal, or how liberal, should I be when given information like this?

"Fairly long life. Into his seventies."

"No, longer. He died at ninety." Then things start to get really interesting.

"Have you done automatic writing?"

"No."

> *"Perhaps you have and not realized it. Alfred's inspired you to do automatic writing. He's already been trying to help you to do that. You may think it's your own ideas you're writing, but you're being inspired."*

June-Elleni is basically reinforcing ideas suggested by other mediums I consulted in my attempt to speak with Wallace – Fransisco in Indonesia, Nasrin and Rita in London, Brigitte in Switzerland – that Wallace is helping me, supporting me, even inspiring me in ways that I might not be aware of. My emotional right brain is thrilled. The question for my ever-present logical left-brain niggles – is she telling me what I want to hear? Something that will make me happy? Or should I accept it in good grace, especially since other mediums have made similar claims? I probably should give Wallace co-authorship for my book.

> *"Stubborn. Wasn't very agreeable. Liked to cooperate but had his own ideas."*

"Yes."

> *"I'm getting a long-suffering wife."*

Another big error. "No, he had a happy marriage."

> Then June-Elleni self-corrects. *"Maybe he had a good marriage, and this was his way of teasing; he could only say this if his marriage was good, tongue in cheek."*

She's good at this self-correction. And she's such a nice person I want to believe her winding U-turns are detours done in good faith and are true to her contact with Wallace. But still, it feels like a psychic-juggling trick.

"Urge to know different cultures and religions and different ways of living in the world."

Now this is interesting. My left-brain tells me that June-Elleni is picking up this concept from *my* interests. Perhaps she uses telepathy (which in itself would be a wonderful, magical gift). Perhaps she picked up clues by looking me up on Google. And when we were on Skype, she could peer into a corner of my study, which features numerous Ganesha statues, shelves of books, old lithographs, and, most prominently from the way my camera is set up, two rare and dramatic African masks.

"He was fascinated by psychic ideas but wasn't sure about whether he could trust his instincts. He hemmed and hawed before declaring himself. But once he was in, he was in."

Okay, where does this come from? I've given her no clues that Wallace was a spiritualist. He probably did prevaricate before coming out of the psychic closet, but once he did, he was a full-fledged believer and jumped through many intellectual hoops to justify why he, a scientist, believed in life after death.

I ask: "Does the name 'Ali' mean anything?"

"Feels familiar. Was Ali some kind of mystic? Ali was a teacher of Wallace?"

If I want to, I can make June-Elleni's Ali-answer make sense. But it would be a stretch. Once again, Ali chooses to remain elusive. "And does the name 'Darwin' mean anything?"

Here June-Elleni shudders. Suddenly. I can see it in her face — surprise, anxiety perhaps. *"All the hairs on my arm just stood up. I have goosebumps all over. Rival. That's the first thought that comes to me. Non-cooperation. Frustration. Hand-wringing frustration. An exchange*

> *of information that was incorrect. Some information deliberately incorrect. Not a lot of cooperation coming from Darwin, he even sent 'red herrings' – put Alfred on a wild goose chase."*

Okay, this is crunch time. Let's give June-Elleni the benefit of the doubt that she has no idea we're talking about Alfred Russel Wallace or that she is aware of Wallace's relationship with Charles Darwin – extraordinarily polite, even warm on the surface, but perhaps riddled with doubt and distrust. Her reaction, via Wallace, is impressive and disturbing. I kick myself; on reflection, I should have given her "Charles" instead of "Darwin."

"Where does his spirit live?"

> June-Elleni starts a long explanation: *"It's different for everybody. Alfred lives in a mindscape, he can choose anything and anywhere that he's experienced, any age, any situation. He seems to know about remote viewing."* She talks about ley lines, time travel, resonance, and vibration. Fascinating but no room to explore in this chapter.

"Our time's limited, can we return to Alfred?"

> *"I see the masks on your wall. They embody energy you don't have. Put a mask on, and you will be imbued with mask-energy. In the West, we hide behind masks, but the shamans will tell you that they wear masks to channel the energy within the mask."*

Yes, there *is* energy in my masks. They're the real deal, used by shamans in villages in the Congo and Gabon, and imbued with woodfire smoke and ancient incantations. We're off the track, though.

> *"A terrible toothache. Two children, a boy and a girl."*

No idea about a toothache, but he had frequent fevers, infections, and other physical ailments. He was a brave explorer but accident-prone. And he had *three* children, two boys and a girl.

"Very private, doesn't like to speak about his personal life."

Could be.

"Had many, many interests, some very obscure. Not a nine-to-five kind of guy."

Absolutely. He was not only a polymath but also a poly-campaigner with a long list of interests; these concerns generated 21 books and some 600 articles and scientific papers.[173]

"Sometimes wanted to escape and stay on a beach."

While I can't imagine Wallace lying on a sun chair on the Costa del Sol, I can easily visualize him sitting in his garden in England and daydreaming that he was back on a tiny, obscure island in Indonesia, isolated, calm, intrigued, eating rice for all his meals, and puzzling over the taxonomy of a new beetle he had collected before breakfast.

"Just sit on the beach, watch the mist on the sea, the phases of the moon."

Now, this is both insightful and frightening. In *An Inordinate Fondness for Beetles*, I imagined a modern-day conversation with Alfred Russel Wallace that took place on a real-life beach on Aru Island. The deserted beach has no official name, but I dubbed it Shiva's Beach, a place of death and revival — a few thousand words that I'm immensely proud of. So, the question in my mind: Has June-Elleni been reading my mind? Are other mediums reading my mind or using some form of emotional telepathy? When a medium senses conflict between Wallace and Darwin, is she somehow picking up emotional cues that I am unwittingly sending?

"A friend. George. Lost a finger on his hand. Alfred relied on George for a bit of sanity, to check his thinking. They shared a kind of black humor."

One of Wallace's best friends was George Silk, and Wallace wrote, from Indonesia, that when he returned to London "there will you be seated on the same chair, at the same table, surrounded by the same account books, and writing up on paper of the same size and colour as when I last beheld you." Wallace's letters illustrate a good friendship but also mark a difference between Wallace's exuberant, transient, insecure lifestyle with the safe and boring existences of his stay-at-home friends. Wallace is perhaps gently saying: *When I return to England, I will have changed; you probably not.* I recall Kipling's comment: "All things considered there are only two kinds of men in the world – those that stay at home and those that do not." I have no idea whether George Silk lost a finger.

And then, some 20 minutes after I first mentioned Ali, June-Elleni invokes him.

"Ali was an alchemist, changing people's minds. There's more to that character than meets the eye; the way he gave information through a back door."

And so June-Elleni repeats the view that Ali was a wise teacher. I don't reply but think that, yes, Ali was a teacher in the big sense of the word. Ali tutored Wallace (and Wallace tutored Ali).

"He is keen to feed you more information via automatic writing."

Unclear whether she is referring to Ali or Wallace.

"That will help you to tap into Ali's character."

This is different to what Nasrin and others have told me, which was basically to leave Ali alone. And June-Elleni comes back to Darwin.

"I'm getting twelve percent. Alfred has a twelve percent trust of Darwin."

A lack of trust is logical and makes sense. But where does the 12 percent figure come from? And then she comes back to her oft-repeated refrain.

"Alfred is your spiritual guide; you will uncover things that have been covered up."

A cover-up? I take this as reference to the argument that Darwin plagiarized Wallace's Theory of Natural Selection. Our time is up. Through it all June-Elleni has been sketching. I ask to see her "automatic" drawing of Alfred. And here is where things get really weird. She holds up her sketch of Wallace. And I am, for lack of a better word, flabbergasted.

In June-Elleni's version (left), which resembles a police sketch, she's got his abundant beard right, also the slightly thinning hair. But I point out to her that she hasn't given Wallace the spectacles he wore throughout his life, as shown in the portrait (right) of him in his mid-40s. She replies that I'm wrong; Alfred *had* told her about the eyeglasses, and I should zoom in. Indeed, they are there — barely visible frameless glasses.

A week later, June Elleni sends me an email.

> She explains a serendipitous conversation she had with Gill, the principal of the College of Psychic Studies.
>
> *"I was at the College and mentioned to her that I did a reading with a man named 'Alfred.' I told her that during the reading he said he was associated with the College. So I showed Gill my drawing. She was stunned and said 'come with me.'"* On the fourth floor of the College, where June-Elleni had never been, the principal showed her a portrait of Alfred Russel Wallace, one of the early members of the College.

I ask June-Elleni if this incident happened before or after I had told her Alfred's full name or given her any information about Wallace's achievements and interest in Spiritualism.

> *"Before you gave me his full name. Definitely before. This was when I only knew him as "Alfred,"* June-Elleni says.

She has other interesting things to report; we Skype again.

> *"I got several visits from Alfred. "He told me, 'This is Alfred, friend of Paul's.'"*

Once again, my easily massaged ego is pleased that Wallace considers me a friend.

> When June-Elleni recounts their conversations, she is as vivacious and exuberant as the first time we spoke. *"Alfred told me to look out my window, 'right now!' I saw an intense blue light on the house behind mine. He told me to take a photo immediately, then another in thirty minutes. At the thirty-minute mark, the blue light disappeared."*

I remind June-Elleni that it's March, and weather changes constantly.

> *"He then told me that at one-fifteen I should have my lunch outside, and the sun would be out for twenty minutes. At the time he said that there was horrible wind and rain. But at one-fifteen precisely the sky cleared, for just twenty minutes."*

Alfred then gave June-Elleni, his new psychic-buddy, a long lesson on metaphysics, modern science and genetics, and the healing power of color. I urge her to get back to the main point – my conversation with Wallace, not hers.

> *"Then he came to me one evening, during the moment between waking and sleep. He said, 'Hello. It's me again.'"*

"He sounds like he's stalking you." I recall June-Elleni mentioning that she had a number of psychic celebrity-pals. She goes on. And on:

> *"The black-and-white triangles I see exist to help me heal. They are related to the ferro-magnetic elements in our sinuses that help us orient ourselves."* She says that Alfred explained about the ferro-magnetic elements in our sinuses that help us orient ourselves, and he gives her the chemical formula Fe_3O_4, which is iron oxide; *"crystals of this chemical in the brain help whales orient to the North Pole."*

I listen politely as she reminds me of her earlier comment about "whales/Wales," which I had thought refers to the nation Wales, but which might indeed have referred to Alfred's new comment about the navigation ability of whales.

She continues along what, for me, is an excessively confusing path.

> *"The blue light the previous day is designed to repair my DNA." "Don't eat grains." "I should use fulvic acid and humic acid to clean my crystal."* [174]

By this time I'm bewildered by all this New Age advice. And I'm also a little jealous that June-Elleni has developed such a close relationship with my mentor. He's giving her quasi-medical advice about how to balance her life. *Hey, what about me? He's my pal!*

> *"Hybrid barriers are there for good reason, in plants, animals, and insects. Alfred is worried that such hybrid barriers are upsetting brain chemistry. Restore balance — take a salt bath." "We don't follow natural law. Hybridization causes acid in the body." "Healing power of colors — red equals warmth, yellow equals safety, blue repairs DNA codes, white relates to harmony."*

This conversation goes on for a while. My earlier cautious appreciation for June-Elleni's mediumship morphs into deep skepticism; the whole experience sounds like a woo-woo parody. But then there's the uncanny psychic drawing . . .

CHAPTER 17

Spirit Conversations with the Elusive Ali

While I had modest success "speaking" with Wallace, I had no success trying to channel Ali. But it sure was fun trying.

Most of my attempts took place in Ternate and other locations in eastern Indonesia. Why? My simplistic brain figured that Ali's spirit would hang around the place where he lived and died, and a local medium would have greater empathy with his spirit than say, a medium from Switzerland. My hedonistic brain said: *Hey, this is a great excuse to travel around eastern Indonesia and get some new experiences.*

Does the spirit of a dead person "live" in geographic proximity to the location where the individual resided while alive? Or does the spirit exist where the individual died? Or is the spirit "everywhere and nowhere," much like a search on the internet for a like-minded colleague on the far side of the globe?

Did Ali believe in ghosts? Almost certainly; it's hard to find someone in Southeast Asia who *doesn't* have a ghost story to tell — sometimes it's personal, sometimes it's of the variety "this really happened to my brother's sister-in-law's hairdresser's boyfriend."[175]

Wallace wrote about the belief that men can turn into crocodiles:[176]

> It is universally believed in Lombock [Lombok] that some men have the power to turn themselves into crocodiles, which they do for the sake of devouring their enemies, and many strange tales are told of such transformations.[177]

Wallace then recounts a discussion among his servants about ghosts; note his sarcastic tone:

> I was therefore rather surprised one evening to hear the following curious fact stated; and as it was not contradicted by any of the persons present, I am inclined to accept it provisionally, as a contribution to the natural history of the island. A Bornean Malay who had been for many years resident here said to Manuel, "One thing is strange in this country – the scarcity of ghosts." "How so?" asked Manuel. "Why, you know," said the Malay, "that in our countries to the westward, if a man dies or is killed, we dare not pass near the place at night, for all sorts of noises are heard, which show that ghosts are about. But here there are numbers of men killed, and their bodies lie unburied in the fields and by the roadside, and yet you can walk by them at night and never hear or see anything at all, which is not the case in our country, as you know very well." "Certainly I do," said Manuel; and so it was settled that ghosts were very scarce, if not altogether unknown in Lombock. I would observe, however, that as the evidence is purely negative, we should be wanting in scientific caution if we accepted this fact as sufficiently well established.[178]

Wallace's spirit was talkative and friendly when I tried to "talk" with him, but Ali's spirit was elusive and seemingly not interested in making contact. "Obviously," as one failed-to-connect medium said to me, "you don't have a strong connection with him."

Nevertheless, the quest was extraordinarily entertaining because I got an inside peek at the modus operandi of mediums in five countries. As they say, "that alone is worth the price of admission."

Part I:
The Three Dukuns

Scratch Indonesia's cosmopolitan surface and you'll find a plethora of shamans, mediums, psychics, and soothsayers, all of whom claim some ability to speak with *hantu*, the Indonesian term for ghosts and spirits.[179]

Ofa Firman and a few other friends recommend I meet Nurdin Amin, commonly known as Om (Uncle) Udin, a noted *dukun* (shaman) whose uncle was related to an earlier sultan of Ternate. *He's very good. Well known. He lives behind the Muslim cemetery. Bring him some cigarettes.*

With my friend Azis Momanda, I easily find his simple yet comfortable house that indeed is just behind the Salero cemetery that adjoins the sultan's palace.

"I'd like to speak with Ali," I say, showing Udin the only extant photo of Ali, taken in Singapore in 1862 when Ali would have been about 22. As with other encounters in Indonesia, I find a request like this is received with the same "sure, give me a minute" response as if I am in a department store and ask whether a particular brand of a home espresso machine comes in red.

Udin, 50, a slender man with an impressive moustache, long thin hands, and large eyes that shift quickly between a warm gaze and an intense stare, opens one of the packs of *kretek* I have brought, places one of the clove cigarettes in a black holder, and takes a few puffs before replying. "I'll need some help with this one."

My friend Azis, who acts as translator, [180] and I walk a few hundred meters to a similarly simple and comfortable house on the main road. This is the home of Ibu (mother, madam) Ratna, a carefully dressed woman of a certain age with a gentle, calm appearance. She wears a black hijab and a bright blue batik dress.

We are soon joined by a third medium, Ibu Dayu, 37, originally from Bali and married to a man from Ternate. Dayu has a small business selling decorative stones from nearby Bacan island. I ask if she deals in magic stones. She laughs. "Just normal jewelry." She shows me some of the rings she has designed.

The three mediums work together, concentrating their efforts toward Ratna, who sometimes asks Dayu or Udin for clarification. I appreciate the idea of having multiple mediums focusing on Ali, and, since I appreciate the power of triumvirates — Musketeers, Tenors, and Stooges — in my mind Ratna, Dayu, and Udin become The Three Dukuns.

> Ratna clearly has the power. She closes her eyes and almost immediately begins coughing. It is a sharp, high-pitched cough. Her hands tremble. She starts crying. Ratna's voice is thin and high-pitched. "*Sad. So sad.*"

She relates Ali's story over the next 15 minutes. It is not a logical story in which incident A is followed by incident B. But why should I anticipate otherwise? Virtually no one tells a story in a clean chronological timeline; this is particularly true when one is chatting with a spirit for whom (I assume) time has little meaning.

> *"Ali had a problem with a girl in Ternate. She broke his heart."*

"*Who* broke his heart?"

> *"A girl. His wife. He doesn't want to talk about it."*

"Ask him to talk about it."

> *"Very sad."*

Ratna is crying.

> *"Ali and Wallace were good friends."*

No points for Ratna; when we first met, I had mentioned Ali and a man named Wallace were friends. Like other mediums, upon hearing that they were "friends" she assumed that Wallace and Ali were on a roughly equal social status.

> *"Ali is a romantic man. He liked scenic places. He liked to take his wife to Akah Rica beach in Ternate where they watched the sunset."*

Here's the classic conundrum. Is Ratna channeling something Ali said, or is she putting her local spin on the conversation by mentioning a scenic place frequented by modern young Ternate lovers that she *thinks* people of Ali's period would have gone to relax?

> *"They liked to go to Akah Rica. So romantic."* Ratu cries again.

If we accept that spirits live in a netherworld unburdened by oceans, distances, and cultures, then why am I intent on speaking with Ali's spirit in Ternate? Couldn't I have similarly channeled Ali with a medium in any country with a medium who speaks any language?

One reason, I suppose, to go to the geographical source when pursuing History by Hantu is to get the benefit of the local belief systems. If I were to consult a medium in, say, Rwanda, the hypothetical Rwandan medium would no doubt see things through a Rwandan filter. I think cultural perceptions are important. In this case we get a vision of Ali's life through a Ternate viewpoint — the local perception of an ideal courtship, which she interprets as romantic sunset walks at Akah Rica beach.

> *"He is not too tall, not too fat."*

That describes perhaps 80 percent of the people in this region.

> Ratna continues to cough. Her arms shake. *"His wife is from the royal*

family of Jailolo."

This is interesting. She's referring to neighboring Halmahera island, where Ali nursed Wallace during his bout with malaria and Wallace wrote the outline of his Theory of Evolution by Natural Selection. Wallace later refined and mailed the essay to Darwin from Ternate, hence its common description as the Ternate Essay.

"She looks Arabic. Her father is Arabic."

I want to shout. "Her name! What's her name?"

"Her name is Shinta Qomariah Kaulana."

This is huge. We have a name for Ali's wife.

"She and Ali have one son: Ahmad Kaolan Ibrahim Djafar."

And a name for their son!

"They lived in Foramadiahi village."

I know the place. It's one of the four original villages of Ternate and an important historical site.

Ratna continues sobbing.

"Ibu Ratna, why are you crying?"

"Wallace asked Ali to accompany him to Singapore."

Loss of credibility for Ratna — this isn't a psychic breakthrough since I had earlier mentioned that Ali and Wallace separated in Singapore and Ali subsequently returned to Ternate. I had given too much away.

"Ali asked Shinta to go with him to Singapore. She refused."

This is news. I wonder what reason she had for refusing? Afraid of travel? Stubborn? Unwilling to endanger her son? Or perhaps she saw this as a chance to escape a loveless marriage? I recall Wallace's contradictory comments in *The Malay Archipelago* about the relationship. He said that Ali's wife "would not let him" travel with Wallace to Ceram, then noted that "his wife lived with her family, and it made no difference in his accompanying me wherever I went till we reached Singapore on my way home."

"When Ali returned to Ternate, he was broken-hearted that Shinta and their son had disappeared."

Shinta abandoned Ali! I wonder. Maybe I should channel Shinta?

"Ali never saw them again."

"What!?"

"They were gone when he returned. She disappeared. Left him. He was heartbroken."

That could explain the crying.

"So sad," Ratna says again. She's fixated on the domestic pain. Ratna regularly loses contact with Ali; it's almost like having a dodgy internet connection or a broken radio signal at the far end of the broadcast range. It's as if Ratna is sitting in the dark on a stormy night, waiting for a burst of lightning to illuminate her surroundings for a brief moment. During the disconnect time, she is patient, sometimes sitting quietly while waiting for the contact to be restored,

occasionally chatting with Dayu. *"Ali died at around the age of forty."*

Big problem with this answer. This contradicts Thomas Barbour's report that in 1907 he met Ali in Ternate, when Ali would have been around 67.

The connection is broken again. Ratna pauses to ask Dayu a question. She re-enters her trance.

"Where is Ali buried?"

"Foramadiahi. He was a follower of Sultan Babullah."

Ratna has gone off-piste. She's referring to Sultan Babullah Datu Shah, one of Ternate's most prominent sultans — the Ternate airport is named after him. He reigned from 1570 to 1583, some 300 years before Ali lived in Ternate.

This is another example of Ratna's cultural bias. She has accorded Ali a noble status that is almost certainly bogus, simply because she (perhaps in good faith) feels it is correct. I am a foreigner, with perceived status myself, and I think she feels that Ali should also have a similarly respectable social position.

Ratna is exhausted, and the connection is broken.

I ask if we can visit Ali's grave.

Azis goes out to hire an *oplet*, a public minivan. The cost is about $20 for half a day. We all pile in and drive half an hour to Foramadiahi village on the other side of the island. The oplet goes partway up a steep, narrow, paved path, then the path becomes too precipitous, and we are forced to walk in the mid-day sun. At the top of the hill, we turn left and walk another 20 minutes to the grave of Sultan Babullah. It's an historic site in Ternate, well-maintained with a gate and signboards at the entrance. The sultan's grave, under a mature banyan tree, is surrounded by half a dozen other graves. Some graves have worn headstones markers, but they offer no hints about the occupants because Muslim graves do not have inscriptions identifying who is buried there. We sit down around a grave that has been renovated with white bathroom tiles within the past decade or so. The grave is prominently placed in the shadow of the banyan tree and just below the grave of Sultan Babullah.

We sit around the grave. Dayu's young daughter is bored and wants her mother to hold her.

> Ratna goes into a trance. *"Sad. Sad."*

I ask why Ali, a commoner, is buried here, next to a great Sultan.

> *"Ali was a* tariqat."

Ratna is referring to a person who has studied a mystical form of Sufism.

But my Western brain, always seeking a logical explanation, still can't figure out why Ali would be buried in this particular site.

> *"He became a holy man and god gave permission for him to be buried here."*

This is unsatisfying. Ofa Firma later confirms that the grave around which we sat is occupied by one of Sultan Babullah's relatives.

We are at the point of diminishing returns. We are all tired. I think we have asked all we can of Ratna.

Azis Momanda

The Three Dukuns (plus Dayu's daughter and the author) at the purported grave of Ali, adjacent to the grave of Sultan Babullah, in Ternate.

Along the coast road we stop and buy a few bottles of water. I give a tip to the mediums and retreat to the comfort of my air-conditioned hotel room to take a headache tablet. I realize I didn't ask the single question that might help us find Ali's descendants – his name. I am skeptical of the shaman-influenced name offered by Kuching-based Tom McLaughlin and Suriani binti Sahari – Ali bin Amit – and I am keen to have other suggestions.

While this particular intervention doesn't result in a big breakthrough, Ibu Dayu says we should try again, that sometimes the signals get crossed. And if that fails, well, there are plenty of other dukuns in Ternate.

On a subsequent visit to Ternate I meet again with The Three Dukuns. They say they have new information that Ali's descendants live in Halmahera. I don't have time to go with them, but Emelya, a guide from the Ternate mayor's office, offers to accompany them. The Three Dukuns say they will need two days, with a budget of a couple of hundred dollars. I'm tempted, but I am uncertain how the results would be communicated to me and would prefer to travel with them myself. However, after I leave Ternate, Emelya continues the negotiations, and the Three Dukuns tell her the search will take about a week, and the budget for their time and services has increased to $900. Through Emelya I tell them not now, maybe when I come back to Ternate.[181]

Part II: The Spirits That Keep Office Hours

I climb into the front seat of the air-conditioned taxi at Ternate's Sultan Babullah airport. After a few words of greeting — *May I know your name? How's life? Will it rain this afternoon? Who won the election for governor?* — I ask the driver, Irjan, if he knows any good mediums.

And that's one of my criteria for choosing which places in the world to visit. Not just the presence of mediums but the reality that I can ask a total stranger for his recommendation of a high-quality medium as easily as asking where I can find the best seafood in town.

With just a moment's hesitation (*Who is this foreigner who just entered my life?*), Irjan replies that his cousin might be suitable. She's famous and works with many of the important people in Ternate, a claim I've heard about most of the Ternate mediums I've met. He'll arrange a meeting; we'll see her the next day.

Amalia is in her 30s, wearing a pink hijab, a black T-shirt, jeans. She speaks in a soft voice. Perfectly ordinary in appearance and demeanor. With a couple of Irjan's curious friends, who are present either because they have some psychic abilities themselves or simply because they are curious, we drive out to a simple seaside recreation area. It's a weekday afternoon, and the place is almost deserted. We sit on benches under a thatched-roof shelter and drink fresh coconut water. I explain that I'm looking for Ali. Can she help?

"We'll do it next Wednesday night. Or Thursday night if you prefer."

Well, the day we meet is a Saturday, and I explain I have a flight out of Ternate on Tuesday.

"Sorry."

I'm not sure if she's busy or suddenly changed her mind.

"The spirits are only available on Wednesday and Thursday nights."

I had heard of mediums not being available due to other engagements or illness or needing time to prepare. But I've never heard of spirits of dead people with strict office hours like a bank or government office. I've learned that in Indonesia it is good to be optimistic, to spread your net widely and hope for the best, but be prepared to be disappointed. This is one of those times when I need to take a deep breath. "Too bad, next time I'll come back on a day when the spirits are working."

Part III: The Medium of Transformation

The most dramatic of my dukun experiences in Ternate occurs when I meet Iis Ariska Abdullah.

As with many of my encounters, this meeting involves some legwork and a bit of serendipity.

Friends told me that when they visited Ternate (also in search of Ali), they employed a taxi driver and guide named Mansur, who had a relative named Iis who was a medium. Unfortunately, they did not keep their phone numbers, full names, or addresses.

After asking other taxi drivers, guides, and hotel staff for several days, I finally locate Mansur, a talkative fellow who enjoys telling me about the important people he has ferried around the island. He is happy to introduce me to his cousin Iis (exact family relationships can be fluid).

Left and center: Paul Spencer Sochaczewski. Right: Rinto Taib.

Left: Iis channeling Ali through her grandmother-spirit guide.

Center: Iis channeling Ali through her cigarette-smoking bearded grandfather.

Right: Iis (in center), out of her trance, with her family and the author.

Ternate-based medium Iis Ariska Abdullah has a remarkable ability to transform herself while in trance. But when not speaking to spirits, she is just a normal young woman who works in a government office, has a young child, and lives a comfortable middle-class life in Ternate.

Iis gives readings to the Ternate royal family and is well-known in Ternate spiritual circles. She is a normal-appearing young woman, works in a government office, and lives in a comfortable, spotless, solid middle-class house. Family photos adorn the walls, the furniture is, to my taste, too heavy and formal for the tropics. I'm welcomed by her proud parents. If they are nervous, it's perhaps because it's unusual for them to welcome a foreigner, not because they are apprehensive about their daughter's upcoming trance.

I explain briefly that I want to speak with Ali and find his descendants. I try not to give too much away.

Iis slips away to her room and returns wrapped in a Muslim prayer shawl, a *mukena*. This garment, a form of chador, is usually a sober white or pastel color. Iis's mukena, however, is muted gold with a motif of embroidered red flowers.

It is as if Iis has donned a magic cloak. She undergoes a transformation as dramatic as any I've seen. The sweet, quiet young woman has become a crone, scrunched and jittery, with a high voice that conveys a feeling of deep sadness.

Iis wiggles her fingers, twists her body to make herself even smaller in the large chair, rocks back and forth, and cries out. Her face is in extreme pain. She looks like she is paralyzed, and she has trouble speaking. Her mouth is twisted, and she chews her words. She later explains she was channeling her grandmother, who would speak with Ali.

She suddenly laughs, not the giggle of a young woman or the response to a joke but a cackle I would imagine coming from one of the witches in *Macbeth*.

Her eyes are closed, then she opens them. But even though she faces me and my friend Rinto, who is acting as the translator, she doesn't seem to see us. Yet she responds to questions.

"You are a traveler," she says to me.

What a disappointing first comment.

> *"You are the most popular man in the world for Ali."*

I think I understand her meaning that I've taken an inordinate interest in the man. Nice, but still an empty bit of praise.

I ask her the usual questions about Ali's background.

> *"Ali is not from Sarawak."* She explains he is from Ternate and was sent to Sarawak as an ambassador for trade and to promote Islam.

He would have been a young boy, I argue.

Iis tries to clarify. Has she been caught in a wrong guess and needs to talk herself out of trouble? But she's clearly in a trance, so I have no idea what's going on in her head. How does she come up with this stuff?

> *"Ali went with older relatives because they sensed he was bright and that travel might be enlightening for the lad. They were all from a "normal" family that was nevertheless related to the sultan [that would have been Sultan Muhammad Zain, who reigned 1823–1859]."*

I subsequently checked with my friend Ofa Firman, a son of the late sultan; he does not believe this story.

This seems to be another example of a medium exaggerating Ali's importance, similar to how Ratna of the Three Dukuns told me that Ali was a holy man and, therefore, deserved to be buried next to Sultan Babullah, Ternate's most famous ruler.

I want to clarify whether Ali was from a noble family or was a commoner, but Iis is rushing ahead, clenching her fists, making whimpering sounds, tightening her body into an awkward posture.

> *"Ali had conflicts with other people. He had a different perspective."*

I have no idea what that means.

> *"You are on a holy mission."*

Nice of her to say, but I disregard it as a culturally biased compliment.

> Iis then gives me a piece of information that appeals to my left-brained quest for data. The name of Ali's wife: Siti Humairah. *"She might be mixed Ternate Malay and Arabic."*

Interesting and perhaps important. Although the name is different, this corresponds to the Three Dukuns who said Ali's wife was Arabic.

> *"Ali and Siti had six children, three girls, three boys."*

And she returns to the story that Ali was sent to Sarawak with older relatives.

> *"The sultan of Ternate sent Ali, who was then a boy, and five adults to Sarawak. In due course Ali met Wallace."*

This opens an entirely new concept of Ali, which relates, with a bit of imagination, to the Three Dukuns' claim that Ali is buried next to Sultan Babullah, in Kampong Foramadiahi, Ternate, because he was a holy man.

But Iis clarifies that Ali wasn't a holy man but a *kapitalao*, the admiral of the sultan's fleet.

This is completely new information and quite unbelievable. But before I can ask for clarification Iis is off on another tangent. She gives me more information than any other medium has.

> *"The mother of Sultan Mudaffar Syah wrote a book about Ali. She dictated it to a scribe named Naida in the Melayu [Malay] language, and it*

exists as a handwritten manuscript." [182]

This is an example of a medium inflating a response. After the conversation with Iis, I check with Firman. The truth is that there *is* a book titled *Naida*, but it was written by a prime minister of a Ternate sultan around 1800 (decades before Ali was born). And Firman says that his grandmother (Sultan Madarfasyah's mother) never wrote a book and no such manuscript exists.

Is Iis lying? Misinformed? Unconcerned about accuracy? Concocting a story that sounds plausible to appease me? All of the above?

Then it gets even more interesting. She returns to the kapitalao story.

> *"Ali returned to Ternate and rose through the ranks to become the admiral of the sultan's fleet, the kapitalao."*

And then Iis tells me something that mirrored what Sultan Hamengku Buwono IX of Yogyakarta said when I asked about his relationship with Kanjeng Ratu Kidul, the Mermaid Queen — *some things cannot be analyzed using Western logic. Accept it. Or not.* Happy coincidence? Is Iis reading my mind? Or is this a maxim repeated by all mediums globally?

> *"We cannot look at Ali using logic. Even though he was a cook, the Sultan felt that Ali was special and sent him to Sarawak and then made him kapitalao."*

Well, she's right about his being a cook. But his career path doesn't compute: She says the young, inexperienced Malay boy has morphed into a proud, high-ranking military officer. Another example of cultural embellishment.

> *"Ali is wearing his military uniform. He looks like a gentleman. He wears*

a gold turban. He introduces himself to me, saying: 'I am Kapitalao Ali.'"

Lovely image, but a lot of new, dramatic, and ridiculous information for me to digest. But I'm intrigued by the image of Ali as a man who liked fancy dress. He seemed proud to be photographed in Singapore in 1862, shortly before Wallace returned to England, wearing a stylish European suit. I can accept the idea of Ali being proud to dress up. The Malays, both women and men, of the archipelago are known to be proud of their personal appearance and take pleasure in dressing elegantly. Also, we know from Wallace's writing that Ali was garrulous and liked to be the center of attention. So the concept of an idealized Ali wearing full military regalia makes sense, but the reality is outlandish. This conversation with Iis took place several years before I learned about the well-dressed Ali Kasut (Chapter 12, "Ali Returned to Sarawak?").

"He looked Arabic."

Well, according to that single photo we have of Ali, he had dark skin and could well have been, at least partially, of Arabic descent.

"He was tall."

And here we come unstuck. Wallace said Ali was short. The 1862 photo shows a young man of about 22 but gives no indication of his height. When Barbour met Ali decades later, he describes him as "wizened," meaning hunched over like an old man. This "tall" description is inaccurate. It's another example of a medium promoting the idea that Ali was "greater" — in all meanings of the term — than he actually was.

"Ali learned cooking from his mother, who worked in the sultan's household."

I wonder, did I mention to Iis that Wallace first hired Ali as a cook?

Then a dramatic transformation, as abrupt as a magician's reveal. Suddenly, Iis's posture changes — she becomes less cramped, more fluid and expansive. Her voice shifts register — it becomes deeper, more masculine. She is now in touch with her grandfather (she explains later), and Iis demands a cigarette. We pass her a lighted clove cigarette, which she puffs hungrily.

A missed question. Does Iis smoke in normal life?

> Iis strokes her chin, as if caressing an imaginary goatee. Her posture becomes less cramped; she seems more contemplative. *"There was a conflict between Ali and other people."*

I recall Ratna's comment that "Ali had conflict with other people. He had a different perspective." I wonder if all of Ternate's mediums regularly get together for coffee and share experiences they've had with curious clients.

What kind of conflict?

> Iis doesn't respond to the direct question. She puffs her cigarette, which she holds like a man. *"Your book has many problems."*

A true mind-reader.

> *"Many different perspectives."*

What does this woman know about writing or my book? But she's right, of course. Problems. Perspectives. I've got them all.

> *"You are on a holy mission."*

Again, a holy mission. A quest. Where does this stuff come from? Is this a generic comment made by Indonesian mediums, like the assurance "Your grandmother loves you and protects you," given by British mediums?

"Ali plays games."

The Indonesian language is less precise than English. And Iis is speaking in standard Bahasa Indonesia mixed with dialect. It's unclear whether she is saying "Ali *was* playing games" (a historical observation) or "Ali *is* playing games" (referring to the current conversation). Either statement requires clarification. Playing games with whom? For what ends? I want to shout. *Come on woman, speak in simple, complete, illustrative, declarative sentences.*

"You have to work for it."

That's a message for me from Iis? Or wisdom offered by Ali?

"He was nothing. But special." And she repeats her earlier comment. *"Ali was nothing special. Sometimes ordinary people are special. Don't use logic."*

And it's over. Iis shakes, grunts, and slowly comes out of her trance, a touch dazed, as if awakening from an afternoon nap in a sweaty place. She leaves the room to remove her mukena and returns as just another normal-appearing young Indonesian woman. She has no idea what just took place. Her parents, who had watched the session, offer us tea and biscuits.

Part IV: Make-Believe Fantasy Statements in London

Nazrin Moazenchi is a London-based medium, recommended by a friend. During a one-hour "conversation" with Wallace I naturally ask about Ali.

I've condensed and truncated the information that Nazrin gave me while we sat on the walking path next to Regent's Canal with additional information she provided a few weeks later.

> *"Ali did not tell me where he was born, but I think he said it was a small village near Sarawak."*

Had I mentioned to her that Ali came from Sarawak? I must have. It's not a location known to many people outside of the region.

> *"He is buried near Sarawak in a little church in a private family section. He showed me the church, but I did not see any name; the triangle shape on the church had a dark blue color with a cross on top."*

Ali was Muslim, and the idea that he is buried in a church (and in a special family plot) is ridiculous.

In a follow-up discussion, Nazrin self-corrects:

> *"I see a small building, with a steep roof and a yellow and red circle on the gable. A weathervane. It's similar to some of the graveyard memorials we find in Persia."*

Well, Nasrin is originally from Iran, so she is clearly putting a cultural spin on this reading.

"Where did he go when he left Wallace?"

> *"When he separated from Mister Wallace, he went to Burma and started his life."*

Ridiculous.

> *"[In Burma] he continued to do his own research."*

A number of mediums have given Ali qualities and expertise he didn't have. They exaggerate his importance, perhaps to please me. If a European arrives suddenly, expressing an interest in a man named Ali, logically, that Ali chap must be important. But Ali wasn't a scientist or a researcher. Nor was he part of a high-level diplomatic delegation, later an admiral serving the sultan, as Iis in Ternate said; nor a holy man, as Ratna suggested; nor a scientist as Moazenchi said. He was a simple, good-natured, helpful, probably illiterate lad.

> *"He exchanged letters with Wallace, but as the time passed this contact got less as he had children and was busy with his own life. He kept those letters in a wooden box. However, he didn't say where they are now."*

Absolutely no evidence that they exchanged letters. Like other mediums, Nasrin is grossly inflating Ali's abilities and career path, suggesting he was literate, a Christian, and a scientist.

> *"Ali is with Wallace in a distant galaxy."*

We're in woo-woo land.

> *"Ali is dark, slim, short."*

Yes. "And his name?"

> *"They're not telling me Ali's full name. They don't say. I have a problem when I ask about the name."*

Hmmm.

> *"Ali is curious to learn. They had a teacher-student relationship. Ali had a dream to learn."*

A "dream to learn?" He might not have expressed it like that, but he was certainly curious in his own way.

> *"Ali says he wants to live his life journey. He says he wanted to do a lot of things."*

Interesting. Here Nasrin is describing an Ali who is a keen student on an interesting quest, not an unsophisticated youth.

> *"Had a wife and children."*

Well, yes, according to Wallace. But I recall during an earlier conversation with Wallace via Nasrin, I asked the same question about whether Ali had a family. At that time Nasrin joked, "Go to the archives."

> *"One of Ali's ancestors was from China. Maybe grandfather."*

Highly unlikely. Ali's children?

> *"Descendants. One, a son? — is in Norway."* She adjusts her story, sticking with Nordic countries but shifting from Norway to Finland. *"One of his children is in Helsinki. He said his family is named something like Parsavalus or Parsavarus. I am not sure."*

First, the use of tenses in conversations with spirits is problematic. Consider the verb "is." Is Ali telling me, via Nasrin, that one of his children was in Helsinki at the time he was alive? Or is he referring to his present-day descendant? And to claim that he has a descendant in Norway or Finland (with a vaguely Finnish-sounding name) is implausible, to be polite. She has given me a narrative that is convoluted and silly. She might have argued that details get muddled when talking with a spirit, and that the concept of time has little meaning when speaking with someone who is dead. But let's speculate. Maybe by "children" she means not a son or daughter, but great-great grandchildren. Maybe one of Ali's distant female descendants married a man from Finland (or Norway). It would have been highly unlikely during Ali's lifetime, but in the 21st century it would not have been overly unusual for such an international, intercultural marriage to take place. But those are all wild assumptions in an effort to smash a square peg into a round hole.

> *"Wallace liked to test him. Because they didn't trust others, they started to talk among themselves."*

I like the idea of Wallace and Ali as co-conspirators.

> *"Ali doesn't want to talk. It's really hard to get information out of him."*

Other mediums have said something similar, noting that while I have an emotional bond with Wallace, I do not have a strong connection to Ali. He can choose whether or not to converse. Is this his decision based on my vibrations? Does the talent of the medium come into play? What about the personality and ability of the medium's spirit guides? So many co-conspirators.

Part V:
Ali Remains Elusive in Geneva

Switzerland-based medium Brigitte Favre, who is also a clinical psychologist, has two conditions: no guarantees when she does works as a medium, and she does not want prior information. She gives me an insightful and lucid reading with Wallace; a few highlights:

Accurate information about which she had no way of knowing:

"Understanding about spiritualism. He fought it for a while because he has a scientific mind and took quite a while to accept it. He wanted proof."

"I feel a very strong man. He would go anywhere if there was something to discover."

"Spent lots of time on boats."

"Enjoyed writing – he wrote journals."

"Link with slavery – not him directly but people he was involved with. He met people who weren't free."

"Learning from people."

"Interested in different races of people."

"Quite tall. Round glasses."

"Studied nature."

"He got sick, some disease, couldn't travel as much as he wanted."

"Tried to change society, laws, and convince people. He didn't succeed as

he wanted, he sees it as a failure."

"Humanist. Even avant-garde. Ideas ahead of his time."

She adds several ego-boosting comments about my relationship with Wallace:

"He knows you tried to speak with him previously."

"You want to rehabilitate something about him. He's excited about that, like you could take some of his work further."

"He is honored by your interest. He gives you a little bow and touches his forelock."

In a second conversation I tell her I specifically want to speak with Ali. She is unable to make contact and stops soon after she begins.

I ask: "Why do you think you can't make the same kind of contact with Ali that you did with Wallace?"

For Brigitte this is obvious. *"You have a strong connection with Wallace, and he has a connection with you. Wallace wanted to speak with you. But you have no real connection with Ali, and vice versa."*

Part VI: Ali, a Seated Skeleton

Barbara Kohler is a well-known German medium. On my behalf a friend asked her to contact Ali. Kohler insisted on having more background to make it easier to connect with him, and my friend inadvertently gave too much away — she explained that Ali was an assistant to Alfred Russel Wallace. Barbara had heard of him, and asked: "Wallace, the explorer?"

There are too many variables in this conversation to give it much credence. Primarily, the following exchange is the way my friend recounted it; I wasn't present. Also, it seems my friend gave Barbara too much pre-séance information. Finally, my friend spoke with Barbara in German, over the phone, and while my friend's English is excellent, no doubt there might be subtleties lost in translation.

> *"Ali was born near the sea, small coastal village on an island, perhaps Sumatra?"*

Well, it was Borneo, but Sumatra's a reasonable guess for a European who isn't familiar with Asian geography.

> *"He was indigenous, grew up in poor conditions, underprivileged."*

Likely true, but this could also be Barbara's cultural bias creeping in.

> *"Alfred Russel Wallace offered Ali a job . . . as a simple mate first, hauling boxes, cooking, then he worked his way up."*

Extraordinary, not the fact that Ali was lugging boxes, but that he was cooking, and most important, "worked his way up." Or, had my friend given it away by telling Barbara that Ali was a cook?

> *"Ali could write; you can find Ali's handwriting on documents!"*

No evidence Ali was literate.

> *"So, he kept lists, catalogued."*

She's giving him more credit than he deserved. There is no evidence Ali was involved in cataloguing or written administration.

> *"He got more and more knowledge."*

Now this is interesting and accurate.

> *"He is extremely loyal, faithful... position of trust... scientific tasks were handed over to him."*

True, although it is an exaggeration that Wallace handed him "scientific tasks." Shooting and skinning birds, yes. Does that count as "scientific"?

> *"Ali travelled with Wallace from given circumstances... it was not intended."*

I don't know what she means.

And then Barbara comes up with a bombshell, one of those unexpected left-hook sucker punches that shake my smug sense of cynicism.

> *"Ali's contribution was a caring support... he nursed Wallace during a serious illness."*

If we accept that Barbara had no prior knowledge of Ali and little knowledge about Wallace, how could she know that Ali nursed Wallace. But her statement is certainly true (Chapter 4, "Ali and Wallace – In Sickness and in Health").

"Ali was paid with silver coins; with it, he supported his family, and his tribe."

Possible.

"Ali had no family of his own, no children!"

This is untrue, at least if we accept Wallace's repeated comments that Ali had a family in Ternate.

"After leaving Wallace, Ali returned to his village, and lived with his brother or cousin . . . in any case blood relation, until his death."

Impossible to verify, but unlikely if we accept the common wisdom that Ali was from Sarawak, then spent the rest of his life in Ternate.

"Ali died rather young, between 40 and 50 years. Although he ate mostly fish, he died due to blood-vessel obstruction and heart disease."

This contradicts Thomas Barbour, who met Ali Wallace in Ternate when Ali would have been about 67.

"No one in the village knew about his merits."

Highly unlikely. From Wallace's notes, Ali was a voluble and keen storyteller, often self-aggrandizing himself when recounting stories of his voyage with Wallace.

Where did he die?

"No one can say, Ali died anonymous, unnoticed."

But then Barbara added:

"I see a seated skeleton in a cave."

Now we're in the fascinating realm of Indiana Jones and gruesome funerary rituals. It's starting to get a bit woo-woo, and I love it.

I'm impressed by Barbara. She got some of Ali's story arc correct, particularly the bit about nursing Wallace. And, to be honest, I appreciate the "seated skeleton" image; I'm a sucker for the ridiculous.

CHAPTER 18

Psychics' Impressions of an Ali Portrait – Mixed Results

I decided to try an experiment. Could a psychologist or psychic tell me about Ali's personality simply by looking at his picture? I consulted three experts – one in London, two in Geneva – for help.

I contacted June-Elleni Laine, the psychic in London who had spoken with Wallace and produced the remarkable psychic drawing of his likeness.

I sent her the only photo we have of Ali. But I did not give her his name, mention any background, or reveal the Wallace connection. Nevertheless, you could argue that her reading of the photo was tainted by our conversations two years earlier. She knew I was on the trail of Wallace, and this foreign-looking young man could be part of Wallace's story.

June-Elleni took the initiative. She put the photo in a sealed envelope and asked six of her students who were learning spirit communications to see if they felt anything about the person whose photo was hidden in the envelope.

Some of the thoughts were ridiculous. Some could be made to fit Ali's situation if I was willing to open my mind and not be too literal. And a few were of the "wow!" variety. Here's a summary of the dozens of student observations:

The ridiculous and random:

- Well-read
- Turquoise and green
- Holy man
- Bartholomy/Jacob/Anton (names)
- Caribbean
- Politics
- Studied to be a doctor

The impossible to verify:

- Music
- Left-handed
- Left-foot limp
- Didn't live the life he wanted
- Cruel father
- Two brothers, one sister
- Heritage suppressed

The plausible:

- Seasickness
- Laid to rest in foreign land (ex Borneo?)
- New beginnings
- Keen gardener
- Liked to draw
- Shy intelligence
- Increased stature
- Learned/knowledge
- "Sent away"
- Rebellious against society's laws
- Wants to be treated better
- Helped others
- Middle class later in life (Ternate?)
- Knew secrets

The disturbingly accurate:

- Navy/sea/seafaring/ships
- Healer (nurse?)
- Pointed the direction (guide?)

But the most remarkable reading came from June-Elleni. True, she knew about my interest in Wallace and had obviously seen the photo. Nevertheless, she said she wasn't aware who it was or to what/whom it referred. Her notes are of the goose-bump variety. Make of this what you will:

- Islam
- Borneo
- Clever
- Elevated his status by association
- Loyal to himself, changing his associations to fit his needs
- Started a family young
- Charming and believable
- Gifted with communication
- Seasickness
- Teenager
- Guide in 1800s

And then the kickers:

- Malay
- Singapore
- Ali
- Wallace's son

While the ideas from June-Elleni and her students focused on relatively straightforward biographic details ("teenager," "seafaring," "healer"), Geneva-based psychic and intuition coach Moïra Salvadore delved more into the psychological and emotional worlds of Ali.

We met at La Vouivre, one of the old-style cafés in Geneva, noted for its comfortable chairs, good pastries, and classical music.

I showed her Ali's photo but did not mention his name or background.

Her thoughts:

- Strange feeling, he's hard to read.
- Old soul.
- Has two personalities, each very different. He could be very gentle but also very cruel. Cruelty to animals? Many children will torment an insect or animal when they're young, but he carried on as an adult, maybe to cast away bad feelings?
- Very intelligent.
- A manipulator. He agrees to things he doesn't like to get what he wants.
- Not well treated by others; they don't believe in his intelligence, and he feels bad about that.
- High potential (genius?) but has trouble handling emotions. Maybe autism or Asperger's?
- A lot of sadness.
- Disgust about humanity — revulsion.
- Receives affection.
- Feels he is abused. He lets himself be abused but feels that by giving in, he can gain power. Nevertheless he manipulates the situation to his advantage.
- Knows it's not good for him, but he goes so far, he cannot come back.

As with other well-intentioned psychics, mediums, shamans (and a few obvious bluffers), Salvadore has proposed a variety of ideas about Ali's personality. Some of them seem more or less accurate, or at least feasible (does cruelty to animals equate to shooting birds for Wallace?). Other statements are of the variety that if you want something to fit, then you can make it fit: Ali's purported manipulative behavior; sadness, feelings of abuse, he accepts unpleasant conditions so he can turn conditions to his advantage.

SECTION VI

Conclusions

CHAPTER 19

Ali: A Life Story Rich in Supposition

Ali's biography is based on a jumble of facts (precious few), supposition (plenty), and information sourced from mediums in several countries (contradictory and unreliable and presented in this book as an amusement rather than as a statement of fact).

My best guesses about Ali's history:

- Ali's family name: unknown.
- Ali's wife's name: unknown.
- Ali's siblings: unknown.
- Ali's children: unknown.
- Ali's birthplace: possibly around Kuching or Santubong, Sarawak.
- Ali's occupation pre-Wallace: possibly worked with Spenser St. John.
- After Wallace departed: Ali likely went to Sarawak, where he was a guide for the Boyle brothers, the young English adventurers who hired Ali to show them Sarawak. He then returned to Ternate where he had a family and where he met Thomas Barbour. He possibly worked as a bird collector and taxidermist for Maarten Dirk van Renesse van Duivenbode.
- Death: died some time after meeting Barbour in 1907 (when

Ali would have been approximately 67 years old).

- Ali's burial site: unknown, possibly on Ternate or neighboring islands.
- Ali's descendants currently live in: unknown, possibly Ternate region.

My best guesses about Ali's relationship with Wallace:

- Wallace was generous in acknowledging Ali's contributions. In *The Malay Archipelago,* he mentions Ali by name no fewer than 42 times.
- Ali's assistance helped Wallace succeed in one of history's (and science's) great adventures.
- Wallace similarly gave Ali the opportunity to have experiences that he likely had never imagined.
- Ali nursed Wallace. Wallace nursed Ali.
- They made each other better.

(NB: The author is on a continuing quest to discover more about Ali and locate his descendants; he welcomes information, speculation, and wild theories. Please contact Paul Sochaczewski via his website: www.sochaczewski.com).

SECTION VII

EPILOGUE

CHAPTER 20

Who Is My Ali? And Who Is Yours?

I've procrastinated writing this last section, simply because I have multiple Alis. I'm unable to choose just one individual.

No need to name names:

- The annoyingly smart girl in primary school who always got better grades than I did, a situation which triggered a sense of unrequited competitiveness.
- The high school English teacher who wrote "Nice job" on one of my essays.
- My high school soccer coach who made me realize that I didn't have to play American football to be seen as a competent athlete.
- The irritating teacher in high school (we had shouting matches) who taught me to type.
- The editor of the university newspaper who made me Cultural Affairs Editor, a position that got me tickets to innumerable opening nights which greatly facilitated my social life.
- The bureaucrat who accepted me into the Peace Corps. He rejected my request for a posting in Botswana, but then posted me to Marudi, Sarawak, an experience that changed my life in unexpected and dramatic ways.
- The Army doctor in Bangkok who had a poster on his wall

of Spiro Agnew strangling the Statue of Liberty. He correctly diagnosed me with a wonky back, making me ineligible for the Vietnam draft.
- The people who hired me at WWF.
- The caring soul who looked after my dying mother while I was two continents away.

I could go on and can happily thank countless people who helped me from behind the scenes:

- Editors who have published my writing (even when it was full of youthful pretension).
- People who have helped me get jobs or introduced me to like-minded people who became friends.
- People who have listened to my advice and surprised me by thanking me years later.

But must an Ali always be a positive force? I suggest that Alis can also work in negative ways:

- I thank those people who have closed doors and rejected me, forcing me to recalibrate and rethink.

You will recall Joseph Campbell's archetypes that feature in the Hero's Journey.

We might consider it outrageous hubris to think that we are heroes on such a grand-sounding life voyage, but the principle remains. Each of us is on an odyssey of discovery, a journey rich with all the emotions, achievements, setbacks, and open-eyed wonder that are the ingredients in everyone's life.

And each of us has an Ali, often several, who have quietly stood by us, opened doors, made a key phone call, introduced us to someone with a good soul, held our hands in times of need, given us a break or sage advice, or simply listened to our dreams.

An Ali can come in many guises. We are aware of some Alis and can express our gratitude to them. But there are countless others who have worked in the shadows of our consciousness.

And the reverse is equally true. Each of us has been an Ali to someone. We might not realize it, but we have had an influence on the life journey of countless people. You might have forgotten to whom you gave a helping hand, but chances are your kindness is engraved in the heart of the recipient of your largess.

I welcome readers' thoughts about your own Wallace-Ali-like connections. To whom do you owe a debt of gratitude for helping you understand the direction of your life journey? And who have you helped explore the world?

So, I'll repeat the toast from the Dedication. We don't know Ali's birthday, so let's raise a glass on January 8 — Alfred Russel Wallace's birthday — which I hereby designate as Global Ali Day. Let us honor people who have helped us on our remarkable personal journeys.

CHAPTER 21

Who Was Ali's Ali?

Ali had never considered himself a hero, at least not in the Joseph Campbell/James Bond/astronaut sense. But at the age of 75, sitting on the small terrace in front of his house in a village a couple of miles outside Ternate town, he played with his two granddaughters and reminisced.

Ali's memory was starting to fade, and that affliction, coupled with a natural tendency to enhance reality, meant he often repeated his tales of adventure. That didn't bother his granddaughters, Siti and Aminah, who loved to listen to him while sitting on his lap. They were attentive as they cuddled the kittens that had been born a few weeks earlier to one of the three calico cats that seemingly owned the premises. Ali's words were soothing, his presence meaningful.

"Ah, there was the tiger in Singapore," Ali said. "Did I ever tell you that story?"

"Tell us again Tete," Aminah said.

So, Ali smiled and repeated the tale of how he had stalked a man-eating tiger in the jungle of Singapore for three nights, sleeping in the branches of trees, following the animal they called "The Striped Killer." How one morning after washing in a stream he saw the animal just a short distance away. "I don't know if he was curious or was about to attack me," Ali said. "I spoke to him in English, because even animals take orders more seriously when they're given in the White Man's language. "Don't be afraid," I said. "But he leaped toward me, and I had an eye-blink time to grab my gun that was lying on the mud, roll on my side, and shoot the beast. When they cut him open, they found jewelry and pieces of clothing of some of the people he had eaten." Siti had heard this story many times; it was part of family folklore. Even for the skeptical people in the village, and there were many, each time Ali told the story, it became more deeply embedded in local legend, until the boundary between fable and fact ceased to exist. "The governor of Singapore, a big White Man with a palace as big as Ternate island, who had a hundred concubines and ate off plates made of gold, sent for me and gave me a medal and shook my hand."

Siti, who was ten and, therefore, wise in her ways, listened carefully, for she was a serious girl. Her Tete was a hero, and she didn't care that the old man embellished his stories. She dreamed of seeing the world and asked him for the umpteenth time to tell about the time he meditated with the Sultan of Jogjakarta in his palace, and the Mermaid Queen appeared to both of them in a vision.

Talking to his granddaughter cost Ali some of the little life-force he had left. But for him it was a great gift. "Siti, my little one, some day you will explore the world beyond Ternate."

Siti didn't see how that would be possible. But she kept the thought in her head every time her grandfather predicted her future. *Keep him talking*, she thought. "You did these things all by yourself, Opa?"

Ali sensed that that this clever young girl would not accept a fantasy story in answer to her question. And what is the responsibility of old age if not to tell the truth to a child?

Ali paused to light another clove cigarette. "Alone? No, no one really travels alone." He closed his eyes, as if scrunching his face would help his memory. "There was my father, Ahmad. You never met him; he taught me to be serious. And my mother, she showed me what plants to use when you're sick. There was the imam of the mosque where I grew up in Sarawak. And my older brother,

Mohammed; you never met him because he died before you were born. We used to go fishing together and play pranks on the girls." Ali coughed. Siti saw blood on his lips and gave him her handkerchief. She put an arm around his frail shoulders, and her sister put her head on Ali's knee so he could caress her like he did with the housecats he so loved. "And when I worked for Tuan Spenser, there was a camp manager, an Indian man who taught me how to serve a white man and to speak some English. But he was cruel, not kind at all." Ali was rambling, but Siti didn't care. His voice was soothing. "Much nicer was the old lady bomoh I met when I first came to Ternate, Ibu Ratna. She liked me, I think. She'd feed me – oh, her rendang was the best – and she'd tell my fortune and remind me how strong I was and how I should believe in myself. And, of course, there was Tuan Wallace..."

Endnotes

Chapter 1
The Hero's Journey

1 Although less known than Wallace, Bates is recognized in scientific circles for his achievements. During his nearly 11 years in Brazil (1848–1862), he collected 14,712 specimens (mostly insects), including some 8,000 which were new to science. Not wishing to see his collection sink into the Atlantic, as Wallace had experienced, Bates sent his collection back to England on three different ships. He recorded his journey in his book *The Naturalist on the River Amazons*. He is best known for developing the concept that later became known as Batesian mimicry – a form of mimicry where a harmless species has evolved to imitate the appearance of an unpalatable or toxic species to keep itself safe from predators.

2 Wallace, Alfred Russel (1852).

3 Wallace, Alfred Russel (1852).

4 Wallace, Alfred Russel (1852).

5 Wallace, Alfred Russel (1852).

6 In *The Malay Archipelago* Wallace writes of visiting a village near Makassar where people had never seen a European. "I excited terror alike in man and beast. Wherever I went, dogs barked, children screamed, women ran away, and men stared as though I were some strange and terrible cannibal or monster ... [Even the buffaloes would] break loose from their halters or tethers, and rush away helter-skelter as if a demon were after them ... If I came suddenly upon a well where women were drawing water or children bathing, a sudden flight was the certain result; which things occurring day after day, were very unpleasant to a person who does not like to be disliked, and who had never been accustomed to be treated as an ogre."

7 The classic Hero's Journey archetypes resonate with the life stories of most of us who wouldn't dare call ourselves "heroes." In addition to the Mentor, we might encounter Minstrels who sing our praises and promote our achievements, Jesters who move the goalposts, and the Nancy Reagans who "just say no," blocking our path and forcing us to look for alternative avenues of satisfaction. In our journeys we encounter Allied Armies and Enemy Forces, a Guardian Angel who keeps us out of excessive mischief, and sometimes a Great Leader who takes us under his or her wing and offers us benediction and riches. But most important is the idea that early in the life-journey story a Campbell-Vogler Hero leaves home to undertake a quest. Sometimes that quest is clear (find the source of the Nile), but often it's a generalized "escape" from a boring or unhappy situation with a resulting search for a vaguely understood "something different" (*Thelma and Louise*). A key feature of the "quest" is a major setback, which, in Wallace's case, might have been when his ship carrying him back to England from Brazil caught fire. And finally, the classic hero's journey, at least in literature and myth, demands a satisfactory denouement. The structure we find most pleasing is a return to the beginning. The hero returns to the situation or place where the journey began, but he has changed, as have the circumstances; this is called the circular structure. Look for it in *The Wizard of Oz*, the *Odyssey*, and the *Ramayana*.

8 Drawhorn, Jerry (2018).

Chapter 2
Slippery History, Sloppy Memory, Creative Conjecture

9 Such "truths" can easily morph into widely held myths and legends; witness Plato's 200-word allegory alluding to Atlantis. For an informative article describing how the myth of the American frontier was created, and later debunked by its author, see Woodard, Colin (2023).

10 This is an example of how direct quotes get muddled. The entertainment industry is rife with misquotes, such as "Play it again, Sam," which was never said by Ingrid Berman in *Casablanca*. In the case of *Dragnet*, the now-common phrase "Just the facts, ma'am" was never uttered in the original 1950s TV series but was quoted in the 1987 parody film

Dragnet starring Dan Akroyd and Tom Hanks. In history, too, we find generally accepted statements that have been shown to be false: "Let them eat cake" (Marie Antoinette). "I cannot tell a lie" (George Washington). "Doctor Livingstone, I presume?" (Henry Morton Stanley).

11 For a contemporary example of an apology for misremembering, here's a quote from Prince Harry, reacting to criticism about the accuracy of passages in his book *Spare*. I'm still trying to make sense of it: "Whatever the cause, my memory is my memory, it does what it does, gathers and curates as it sees fit, and there's just as much truth in what I remember and how I remember it as there is in so-called objective facts."

12 Get enough people to believe your story and you'll be on the fast track to creating a religion, an empire, a legend. Do you care about your legacy? If you are a contemporary famous person, you will leave behind a trove of books, letters, self-serving Facebook posts, sometimes embarrassing Instagram photos, interviews, and speeches. Historians can interview your spouses, relatives, friends, and enemies. But what if you are an average, run-of-the-mill, more-or-less "normal" person? My suggestion is to create a paper trail. I had a friend who was a serious, somewhat-successful, moderately well-known conservationist. He was dying of cancer and wished to have his life accurately reported. So, he wrote his own obituary, which he circulated to friends and media. It was accurate, and, if it was a bit self-serving, well, I can understand that. It was heavy on his professional achievements and light on his personal life. His choice. The message is simply this: You can control your legacy. Leave authoritative-sounding documents. Write your memoirs. Create your own Wikipedia page. Review your life through your own lens. Don't leave that responsibility to someone else (even someone who loves you) because that chronicler will filter your life through his or her own memories and value system. You may well choose to enhance, lie, and fabricate – you wouldn't be the first. Warning: Unless you are an all-powerful emperor, you might want to avoid the type of excessive, auto-bombastic self-adulation favored by Assyrian king Ashurnasirpal II (883–859 BCE), who instructed this message to be inscribed on his tomb: "I am important, I am magnificent." In absentia he then reminded his grieving subjects of his military career: "With [my enemies'] blood I dyed the mountain red as red wool, while the ravines and torrents of the mountain swallowed the rest of them."

13 As we know too well, modern politicians are among the most deceitful of people. Starting with the 2023 saga of Serial Liar Congressman George Santos, Matthew Stevenson, in his article "The Great Santos," offers a fine review of the duplicity and mendacity of American politicians, from George Washington to Andrew Jackson to Joe Biden to Barack Obama, to Bill and Hillary Clinton, to, of course, Donald Trump. Call it embellishment, enhancement, or electioneering, it seems that a greater number of important people than we would like lie about something. Who would have thought?

Chapter 3
Everybody Needs Somebody

14 Most accounts by explorers and travelers treat local guides as anonymous helpers. A 2022 *Smithsonian* article about the search for Wallace's giant bee in Halmahera, Indonesia, described the group's local assistant in nameless generic terms: "A local guide then climbed up for a look, Wyman says, made a hand gesture that resembled an antennae and quickly helped build a platform from branches and vines to enable the group to view." But Phoebe Weston, in an article in the *Guardian* about the same expedition, not only names the individual but took the trouble to interview him so he could explain his role in the discovery: "The small team searched locations based on local information and accounts of previous discoverers. 'It didn't seem we'd get a result, but on the way home one day, I saw a nest by chance and studied it with my binoculars,' recalls one of the team, local guide Iswan Majuud. 'I climbed the tree, and used the torch on my phone to see what was in the hole. I saw something move and jumped down, afraid it was a snake. After catching my breath, I climbed again, and told the team the hole looked wet and sticky. I also saw a black and white creature move inside. They concluded this was a giant bee.'"

15 Wallace, Alfred Russel (1905).

16 Wyhe, John van (2018).
17 Wallace, Alfred Russel (1869).
18 Wyhe, John van (2018)
19 Cranbrook, Earl of and Marshall, Adrian (2014).
20 Cranbrook, Earl of and Marshall, Adrian (2014).
21 Wallace, Alfred Russel (1869).
22 Wallace, Alfred Russel (1869).
23 Letter to his mother. Mary Ann Wallace, from Singapore, April 30, 1854. In: Marchant, James (1916).
24 Letter to his mother, Mary Ann Wallace, from Singapore. September 30, 1854. In: Wyhe, John van, and Rookmaaker, Kees (2013).
25 Letter to his sister, Mrs. Sims from Sarawak, June 25, 1855. In: Marchant, James (1916).
26 Letter to his sister, Mrs. Sims from Sarawak, June 25, 1855. In: Marchant, James (1916).
27 Wallace, Alfred Russel (1869).
28 Letter to his sister, Mrs. Sims, and brother-in-law Frances Sims from Singapore, February 20, 1855. In: Wyhe, John van, and Rookmaaker, Kees (2013).
29 Wyhe, John van and Drawhorn, Gerrell M. (2015).
30 Wallace, Alfred Russel (1869).
31 Wallace, Alfred Russel (1869).
32 Wallace, Alfred Russel (1869).

CHAPTER 4
Ali and Wallace — In Sickness and in Health

33 Wallace, Alfred Russel (1869).
34 Wallace, Alfred Russel (1869).
35 Wallace, Alfred Russel (1869).
36 Wallace, Alfred Russel (1869).
37 Wallace, Alfred Russel (1869).
38 Wallace, Alfred Russel (1869). This was not the first time he wrote about injuring his ankle. The first mention occurred while he was in Sarawak, before he met Ali. He wrote: "I had the misfortune to slip among some fallen trees and hurt my ankle, and not being careful enough at first, it became a severe, inflamed ulcer, which would not heal, and kept me a prisoner in the house the whole of July and part of August."
39 Wallace, Alfred Russel (1869).
40 Wallace, Alfred Russel (1869).
41 Wallace, Alfred Russel (1869).
42 Wallace, Alfred Russel (1869).
43 Wallace, Alfred Russel (1869).
44 Prior to his famous malaria attack while in Halmahera, Wallace wrote that he suffered malaria attacks while in Brazil. Then, sometime between July and September 1854, while in Malacca, he was treated by a government doctor with quinine for another attack.
45 Wallace, Alfred Russel (1905).
46 Wallace, Alfred Russel (1905).
47 Sochaczewski, Paul Spencer, and Hallmark, David (2005).
48 For more on the Wallace-Darwin priority debate, see Beddall, Barbara G. (1988), Brackman, Arnold A. (1980), and Brooks, John L. (1984).
49 One point that is often incorrectly stated, by both researchers and the popular writers who regurgitate previously published reports, is the false idea (a slippery truth, repeated regularly, takes on the patina of truth) that Darwin and Wallace submitted a joint paper to The Linnean Society in 1858. True, their individual contributions were conjoined and published with a single title — "On the Tendency of Species to form Varieties; and on the Perpetuation of Varieties and Species by Natural Means of Selection" — and it was published giving both their names as co-authors. But the reality is *there was no joint paper* — they were separate, unequal submissions. Although size doesn't always matter, it is instructive to consider the length of the three contributions. Darwin's contributions were an excerpt of an unpublished 1844 essay (approximately 2,000 words) and an unpublished 1857 letter

to American botanist Asa Gray (1,300 words). These documents, which Darwin admitted were incomplete, showed that he had been working on the question of evolution, but his conclusions were still basic compared to the detailed explanation he provided in his book *On the Origin of Species*, which came out some 16 months later. Wallace's paper, on the other hand, titled "On the Tendency of Varieties to Depart Indefinitely from the Original Type" (the Ternate Paper or Ternate Letter, 4,300 words) described a concept he called "a struggle for existence" that was later termed, by Herbert Spencer, "survival of the fittest." Had Wallace sent it to an academic journal instead of to Darwin, it would likely have been published.

Scholars enjoy exploring the many puzzles surrounding The Linnean Society event: When did the Dutch mail service leave Ternate with Wallace's letter? When did the British post office deliver it to Darwin? How did Wallace's writing influence Darwin's thinking? Did Joseph Hooker's wife, Frances Harriet Hooker, who transcribed Darwin's handwritten manuscript, edit his thoughts? How devious – or how gentlemanly – were Darwin, Lyell, and Hooker in the episode? What happened to Wallace's original handwritten Ternate Paper? All intriguing questions, the stuff for countless PhD dissertations and journal articles.

My interest is psychological and more oriented toward the initial choice that Wallace must have debated with himself (for there was no one in his circle of friends in Ternate who could have offered useful advice). *Should I send the paper directly to one of the scientific journals that have published my many articles previously, thereby ensuring my priority to the theory? Or should I send it to Darwin and ask him to send it to Lyell and hope that they will see what a bright guy I am and welcome me, a poor, auto-didactic beetle collector, into the pure-oxygen world of the British scientific establishment?* For a helpful summary of the co-authorship question, see Smith, Charles H. (2022). He concludes: "What we should call the Darwin-Wallace contribution is not clear, but it certainly is not a 'co'-anything."

50 Barbour, Thomas (1943).

Chapter 5
Ali's Career Path – Camp Manager

51 Wallace, Alfred Russel (1905).

52 Wallace, Alfred Russel (1905).

53 Letter to Richard Spruce, September 19, 1852. In Wallace, Alfred Russel (1905). This simple statement conceals a number of interesting points. True, Wallace lost most of his notes and many of his specimens in the ship fire (but not all – some specimens had been sent back previously and had arrived safely in England). The remarkable fact is that he still managed to write six academic papers (including "On the Monkeys of the Amazon") and two books about his four years in Brazil: *Palm Trees of the Amazon and Their Uses* and *A Narrative of Travels on the Amazon and Rio Negro*. These are not his best books, but they are important ones, and it is remarkable that he was able to write them at all. Lastly, Wallace had hoped that his collection, and the accompanying books and monographs he would write, would have facilitated his access to the British scientific establishment. But without much supporting evidence to enhance his bona fides, he realized that he had to make another life-changing expedition to consolidate his legacy. He considered going to the Andes or east Africa, but finally decided on travelling to Southeast Asia.

54 Letter to Richard Spruce, September 19, 1852. In Wallace, Alfred Russel (1905).

55 Letter to Richard Spruce, September 19, 1852. In Wallace, Alfred Russel (1905).

56 Ironically, bad luck with ships led to one of Wallace's greatest discoveries. In January 1856 he traveled from Sarawak to Singapore, intending to board a ship that was scheduled to sail from Singapore to Macassar (Makassar, Sulawesi). But he arrived late and missed the connection by several hours and was forced to spend four months in Singapore. He then traveled on a different boat with a different itinerary – the barque *Kembang Djepoon* ("Flower of Japan"), which was sailing to Bali and Lombok. During a three-day stopover on the north coast of Bali he noticed that the avian diversity on the island was similar to that of western Indonesia and the Malay peninsula. However, on the eastern side of the deep 25-mile channel separating the two islands, he observed that the bird life in Lombok was

dramatically different to that of Bali. While he wasn't the first to document this disparity, he was the first naturalist to elaborate on the differences and offer a theory as to why the dramatic contrasts exist. He wrote that the fauna of "the western and eastern islands of the Archipelago, as here divided, belong to regions more distinct and contrasted than any other of the great zoological divisions of the globe. South America and Africa, separated by the Atlantic, do not differ so widely as Asia and Australia." This line of demarcation, which extends north through a transitional region now called Wallacea, has been named the Wallace Line. For more on the Wallace Line see Wallace, Alfred Russel (1859c) and Wallace, Alfred Russel (1876b).

57 Wallace, Alfred Russel (1869).
58 Wallace, Alfred Russel (1869).
59 Wallace, Alfred Russel (1869).
60 Wallace, Alfred Russel (1869).
61 Wallace, Alfred Russel (1869).
62 John Bastin, Introduction to 1986 edition of *The Malay Archipelago*, Oxford: Oxford University Press.
63 Wallace, Alfred Russel (1869).
64 Wallace, Alfred Russel (1869).
65 Wallace, Alfred Russel (1869).
66 Wallace, Alfred Russel (1869).

CHAPTER 6
Ali's Career Path — Bird Hunter, Taxidermist

67 Wyhe, John van, and Drawhorn, Gerrell M. (2015).
68 Beccaloni, George (2018). Beccaloni adds that Wallace's taxonomic haul makes "him one of the most prolific describers of bird species of all time."
69 Wallace, Alfred Russel (1869).
70 Wallace, Alfred Russel (1869).
71 Wallace, Alfred Russel (1869).
72 The fascinating theme of how Wallace (and other 19th-century field naturalists) caught, identified, embalmed, skinned, mounted, and shipped natural history specimens is seldom described in literature. Basic bird taxidermy hasn't changed too much since Wallace's day, notes Hein van Grouw, senior curator of the Bird Group at the Natural History Museum in London. He says the main difference is probably "the amount of insecticides, mainly arsenic, as that is not used anymore." One standard guide to field taxidermy of birds is *Birds: Instructions for Collectors* (1970). Along with helpful illustrations, this taxidermy instruction manual lists 13 pieces of "essential minimum" instruments and equipment that are required, as well as 14 pieces of "useful additions." It is unclear how many of these gadgets Wallace carried (not to mention collecting nets, guns and ammunition, traps, mounting equipment, storage boxes, and the like), but we can confidently predict such equipment would add another box or two of belongings to the baggage he lugged around when he moved camp.
73 Wallace, Alfred Russel (1869).
74 Wallace, Alfred Russel (1869).
75 Had Ely and Yos lived on the coast, their products for sale would have been sea turtles (for the flesh and shells), mother-of-pearl, sea pearls, sea cucumbers, and shark fins. As in Wallace's day, much of Aru's commerce goes through ethnic Chinese middlemen, and many of the finished products are enjoyed by Chinese customers.
76 For more about my travels in Aru and discussions about trade in protected species throughout Indonesia, see *An Inordinate Fondness for Beetles*.

CHAPTER 7
Wallace, Ali, and the Search for Birds-of-Paradise

77 Wallace, Alfred Russel (1856b).
78 Wallace, Alfred R. (1869).
79 Wallace's full description of the bird from *The Malay Archipelago* reflects his interest in

recording the scientific importance of the find as well as the monetary value: "The general plumage is very sober, being a pure ashy olive, with a purplish tinge on the back; the crown of the head is beautifully glossed with pale metallic violet, and the feathers of the front extend as much over the beak as in most of the family. The neck and breast are scaled with fine metallic green, and the feathers on the lower part are elongated on each side, so as to form a two-pointed gorget, which can be folded beneath the wings, or partially erected and spread out in the same way as the side plumes of most of the birds of paradise. The four long white plumes which give the bird its altogether unique character, spring from little tubercles close to the upper edge of the shoulder or bend of the wing; they are narrow, gently curved, and equally webbed on both sides, of a pure creamy white colour. They are about six inches long, equally the wing, and can be raised at right angles to it, or laid along the body at the pleasure of the bird. The bill is horn colour, the legs yellow, and the iris pale olive."
And here's the description of the bird from the Cornell Lab of Ornithology, a valuable resource about all the birds-of-paradise: "A large, front-heavy brown bird with a long, pale downcurved bill and a distinctive elongated head with a flat crown and a forehead tuft. Orange legs. Unmistakable males have a green-blue breast shield and long creamy-white plumes extending from the shoulders. Females are entirely earth-brown, lacking the shoulder extensions and shield. Immature resembles female. Found in canopy of tall forests in lowlands and foothills, confined to three islands in north Moluccas. Regularly sighted at known display sites. Very loud when displaying, giving raucous 'WAA-WAA...' and 'KEE-KEE...' notes."

80 Beccaloni, George (2018).
81 Wallace, Alfred Russel (1905).
82 Letter to Samuel Stevens, from Batchian. October 29, 1858. In Wyhe, John van, and Rookmaaker, Kees, editors (2013).
83 The 1863 issue of general ornithology, *The Ibis*, included a price list of "Birds from the eastern islands of the Malay Archipelago" that were for sale by Samuel Stevens and other dealers. The offering included *Semioptera wallacei* at 200 shillings per pair. So, at 100 shillings each (at 20 shillings to the pound, that would be five pounds), what would they be worth today? One inflation calculator tells that that £5 in 1863 is worth precisely £803.75 in 2023.
84 Letter to Samuel Stevens, from Batchian. October 29, 1858. In Wyhe, John van, and Rookmaaker, Kees, editors (2013).
85 Letter to Samuel Stevens, from Batchian. October 29, 1858. In Wyhe, John van, and Rookmaaker, Kees, editors (2013).
86 Wallace, Alfred Russel (1869).
87 Letter to John Gould. Wallace, Alfred Russel (1859a).
88 Wallace, Alfred Russel (1869).
89 Wallace, Alfred Russel (1869).
90 Wallace, Alfred Russel (1869).
91 Wallace, Alfred Russel (1869).
92 Dickinson, Edward C. and Christidis, Les, editors (2014). Taxonomy can be fluid. A quick internet search indicates that the number of bird-of-paradise species ranges from more than three dozen to 39, 42, 44, 45, and around 50. Dickinson and Christidis offer the statistic I've cited: 41 different species divided into 13 genera.

Chapter 9
Ali's Origin – Sarawak?

93 Wallace, Alfred Russel (1869).
94 Wallace, Alfred Russel (1867).
95 Wyhe, John van, Drawhorn, Gerrell M. (2015).
96 Drawhorn, Jerry (2022). Personal correspondence.
97 Drawhorn, Gerrell M. (2017).
98 St. John, Spenser (1863). His book is a wonderful narrative, full of adventure (pirates!), internecine politics in various royal families, pesky rebels, Chinese secret societies, intrigue and torture, colonial derring-do, and hardships (bravely conquered).

99 St. John gave Ahtan a considerable amount of attention in his memoir: Ahtan managed scarce food reserves and dwindling rice stocks, and collected edible native plants during an impending famine. He recorded some of the trivial details that so-enthralled British diarists of the time (he counted the number of deer horns in a longhouse – "forty-three used as pegs"). He was a wonderful cook; St. John wrote: "A most satisfactory savour rose to my nostrils. I found that Ahtan, having discovered a jar of pork fat, was preparing some cakes. I divided them, but he said, 'No, you, sir, have the larger body, therefore should have the larger share.' But the mutual-admiration society eventually came to an end. St. John finally wrote, "[Readers] may be interested in the fate of my boy Ahtan, and I am sorry to say his conduct ultimately made me lose all interest in him ... [he joined a Chinese secret society], called there a Hué, a branch of the Tienti, or Heaven and Earth Society." Ahtan then poisoned St. John with laudanum (a 10 percent solution of opium powder in alcohol) putting St. John in "a stupefied sleep for thirteen hours." While St. John was unconscious Ahtan and another servant stole St. John's "heavy iron chest." The sultan learned of the incident, tortured Ahtan and was prepared to execute him, and only St John's "English ideas of justice" convinced the sultan to spare the young man's life. St. John wrote "Ahtan ... was known to have been a favourite servant, though his conduct was very bad, particularly in dosing me with opium, yet I could not forget his kindness to me during our wanderings in the interior, and asked for his liberty on that plea."

100 Drawhorn, Jerry (2023).

101 McLaughlin, Tom, and Suriani binti Sahari (2017).

102 According to George Beccaloni, the only Wallace specimen in the Sarawak Museum is a dung beetle. Beccaloni, George (2019b).

Chapter 10
Ali's Origin – Ternate?

103 Cranbrook, Earl of, and Marshall, Adrian (2014).

104 Cranbrook, Earl of, and Marshall, Adrian (2014).

105 Wallace, Alfred Russel (1869).

106 For a detailed analysis of the linguistics behind this phrase and a discussion of the various forms of Malay used in the archipelago during the time of Wallace's visit, see: Drawhorn, Jerry (2016).

107 Cranbrook, Earl of, and Marshall, Adrian (2014).

108 The "word lists" in the Appendix of *The Malay Archipelago* include a section Wallace described as "The common colloquial Malay as spoken in Singapore." Drawhorn notes "Singaporean Malay formed the structure of Wallace's lexical comparisons to other languages and dialects, and clearly represented the language he would have studied."

109 Drawhorn, Jerry (2016).

110 Drawhorn, Jerry (2016).

Chapter 11
Ali: Rich, Facing New Adventures

111 Wallace, Alfred Russel (1905).

112 What I find interesting is that people seem to arrive at conclusions in one of two ways. They might have an opinion, then seek evidence to support their views (for instance, people who espouse creationism and use Biblical references and circular logic as their proof). Or they might examine data and come to a conclusion based on statistics, clearly thought-out arguments, and replicable experiments (for example, people who believe in climate change because the science is, in their minds, sufficiently convincing, and they cite the ample evidence worldwide – melting glaciers, for instance). Let's call it the left-brain option (logic, science, mathematics, proven facts) vs. the right-brain alternative (emotion, mystery, gossip, imagination, metaphysics). The balance between these two options is based on our personalities, education level, and social environment. Consider how we judge a politician running for elected office. Do we believe the countless gossipy rumors circulating about her? Is there any unassailable "truth"? For example, do we believe, in our

hearts, that President Barack Obama was born in Kenya in spite of hard evidence that he was born in Hawaii? Do we believe that U.S. astronauts really landed on the moon, or are we convinced the 1969 moon landing was a hoax produced in a top-secret government TV studio in Nevada? Do we believe that every dose of the Covid-19 vaccine contains a tiny metallic chip by which the government can monitor our every move? What is the tipping point that causes our belief to flip? So, it is with the search for Ali's retirement home – some people say Sarawak, most others vote for Ternate. For me, the left-brain evidence is substantial – Ali "retired" to Ternate.

Chapter 12
Ali Returned to Ternate?

113 My novels *Redheads* and *EarthLove* explore this dynamic.

114 McLaughlin, Tom, and Suriani binti Sahari (2017).

115 Drawhorn, Jerry (2023) writes: Ali "almost certainly" outfitted a collecting trip for Charles Allen in 1860. "Ali was still collecting throughout this period and there are only about four to five months that are not well documented. But clearly, this was only a time when Ali was not traveling far from Ternate, probably due to his marriage and perhaps while recovering from the intensive and debilitating expedition into the interior of New Guinea. Brief collecting trips were not, apparently, too onerous, and were too lucrative to turn down. [In any case] this was a time when sea voyages required several months and depended on the monsoon cycle. Six months were average travel times for round trips between the Eastern and Western Archipelago. I think he's barking up the wrong Ali tree.

116 Ali's success as an independent outfitter and collector argues in favor of the likelihood that after Wallace left, Ali easily found work with, first, the Boyle brothers, then as an employee or contractor to Maarten Dirk van Renesse van Duivenbode in Ternate.

117 I asked McLaughlin to explain the contradiction of Ali having a Sarawak wife, since Wallace twice mentions that Ali married in Ternate. McLaughlin suggested that I was looking at the situation with an overly Western perspective in which monogamy is the norm. He noted that polygamy was permitted both by Islamic Shariah law as well as by *adat*, the traditional culture and value system to which Ali would have adhered. Also, according to McLaughlin, Ali had made a promise to look after the children of his brother Panglima Osman (Seman), and familial responsibility in Sarawak overrode the obligations of his marriage in Ternate. To me, it feels like they are stuck in a self-perpetuating loop, starting off with one assumption, then jumping through hoops to confirm each subsequent rationalization.

118 The statement "He didn't even have to think about it ..." is from a personal interview with Tom McLaughlin.

119 Drawhorn, Jerry (2016).

120 In Drawhorn, Jerry (2018), he writes: "There are just too many convergences. The fact that Ali Kasut (and Wallace's Ali) were in Singapore at the same time. That Ali Kasut (unusually) wore Western clothing (and that Wallace's Ali purchased a set of Western clothing ...and not at Wallace's behest, who thought that he looked better in his traditional attire). That Ali Kasut spoke some English. That Ali Kasut was experienced in outfitting the needs of Europeans on expeditions. That Ali Kasut was 'well-travelled' and was familiar with Sarawak and Borneo (they hired him as a guide). That would suggest that he was not a Singapore Malay... most of the trade at that time was undertaken by Sarawak Malays. Ali Kasut seemed familiar with the Brooke administrators. Like Wallace's Ali, Ali Kasut seemed quite responsible and organized.

121 Boyle, Frederick (1876).

122 Boyle, Frederick (1876).

123 Boyle, Frederick (1876).

124 Boyle, Frederick (1876). Ali's chauvinism for Malay fashion contradicts his earlier behavior when in Singapore he purchased, and insisted on being photographed, in the clothes of a Western gentleman.

125 Temenggong is an old Malay and Javanese noble title of nobility (but it also refers to supreme chiefs among some of the tribal groups in Sarawak). The Temenggong of Johor at the time of Boyle's visit was powerful, ambitious, and involved in political intrigue.
126 Boyle, Frederick (1876).
127 In *Adventures Among the Dyaks of Borneo,* the Boyle brothers come across as entitled, well connected, wealthy, curious, rather superficial, and charming young men from good families and Oxford educations. They were greeted in Sarawak with generous hospitality by the British community (White Rajah James Brooke was known for taking in strays, but the hospitality they received from many colonial officials living in the territory at the time was an indication of the Boyles' charm and social status. Their main activities focused on hunting (anything, really, including deer, orangutans, proboscis monkeys, hornbills, pigeons, and curlews), recounting local history told to them by British officials, admiring the women, and offering politically incorrect pronouncements about the habits and lifestyles of the natives of Sarawak. They make no pretense of undertaking a serious venture – "we regarded ourselves as mere wanderers, neither scientific nor anthropological" – and there is nary a Latin binominal, as populates Wallace's books, in sight. The book is funny, observant, snarky ("the old gentleman high up in the civil Service, who guarded his pretty daughter like an Argus, making up in ferocity for his very ordinary vision; [and] the nervous lady, suffering from cockroaches in the imagination." They were free with travel advice for similarly placed adventurers; their entitlement flares brightly when, on the very last page, the lads "recommend most strongly to any reader meditating an Eastern journey (beyond Suez), whether for business or pleasure, to take the French line of steamers at any inconvenience, rather than place himself in the hands of the [British] Peninsular and Oriental Company... I never met with an Englishman who had experienced the two systems, who did not readily admit, at some sacrifice of national pride, that the French line of steamers was preferable in every respect to that which it so dangerously rivals." The lads were serial adventurers, and they made similar trips (and Frederick wrote similar books) about travels in Nicaragua, Haiti, Costa Rica, and South Africa.
128 A few dates are specifically mentioned, and several other dates can be inferred. The Boyle brothers attended James Brooke's farewell party in Kuching just before the rajah left Sarawak on September 24, 1863. And Boyle writes about receiving letters and newspapers via a German trader while in an out-station and "discussing Lee's retreat from Washington." His reference to "Lee's retreat" was probably General Robert E. Lee's famous defeat at Gettysburg, Pennsylvania (Lee never attacked Washington, D.C.), which took place from July 1–3, 1863. Given the time for a newspaper account of the battle to travel from the States to UK, and subsequent mail service to the Boyles' location, we can suggest the date of this citation was around September–October 1863.
129 Drawhorn, Jerry (2016), noted these abilities of Ali Kasut, along with his preference for Western clothing (the "almost smoking gun"), and his debilitating illnesses.
130 Boyle, Frederick (1865)
131 Boyle, Frederick (1865)
132 Boyle, Frederick (1865). It wouldn't be a convincing colonial narrative without occasionally being snippy about the efficiency of the servants.
133 Boyle, Frederick (1865). In contrast with their satisfaction with Ali Kasut, Boyle gleefully tells of his annoyance with a servant named Paham, using terms such as "iodiocy," "vain and indolent," "gorgeously attired," and "lying" (Paham claimed his father was a great chief and he had "fled from home for the murder of a Dutchman"). Boyle's comments are in a similar vein, but less extravagantly written, to those Wallace wrote about his gambling-addicted servant Baderoon (Chapter 3, "Everybody Needs Somebody").
134 Boyle, Frederick (1865)
135 Boyle, Frederick (1865)
136 Boyle, Frederick (1865)
137 Frederick Boyle (1841–1914) himself had a remarkable career. He was a barrister, and, in spite of his protests that he was a "mere wanderer," he took an interest in history, and in 1866 donated a large number of archeological artifacts to the British Museum that he had collected while travelling in Nicaragua. He was a newspaper correspondent in the

Russo-Turkish War (1877–1878), wrote regularly for prominent magazines and journals, published numerous travel books and novels, and, later in life wrote several books about orchids. He committed suicide in 1914.

138 Boyle, Frederick (1865)

139 Drawhorn, J. (2016).

140 Nobody can accuse Frederick Boyle of writer's block; his book, *Adventures Among the Dyaks of Borneo*, was published in 1865, just two years after his departure from Sarawak, while Wallace's *The Malay Archipelago*, which made him famous, was not published until 1869, some seven years after Wallace left Asia.

Chapter 13
Ali Returned to Sarawak?

141 Wallace, Alfred Russel (1869).

142 Wallace, Alfred Russel (1869).

143 Wallace, Alfred Russel (1859b).

144 Wallace, Alfred Russel (1905).

145 Ali had advantages that would have made him a desirable husband. Besides having some cash and a reputation for having travelled widely with a European, Ali likely was acquainted with a Dutchman named Maarten Dirk van Renesse van Duivenbode (whose name Wallace wrote as Duivenboden). Wallace said he was known as the "king of Ternate" in recognition of the man's wealth and influence. Duivenbode was likely to have known of the young man's reliability and experience.

146 Cranbrook, Earl of, and Marshall, Adrian (2014).

147 The wood mulch-slider, or Mueller's three-toed lerista (*Lerista muelleri*), is a small terrestrial lizard, about the length of an outstretched hand. A key distinguishing feature of the species is the presence of three digits on each of the forelimbs and hindlimbs. The genus *Lerista* was revised by Storr, Glen M. (1971), who determined the entire genus is endemic to Australia, with *L. muelleri* restricted to western Australia. Therefore, we do not know what specific critter Barbour and Ali discussed.

148 Barbour, Thomas (1912).

149 Barbour, Thomas (1921).

150 Thomas Barbour was a leading naturalist, with a global reputation. Although he collected all manner of birds and insects (particularly butterflies), he is primarily known for his study of amphibians and reptiles; some 18 new species he discovered are named after him. He led major expeditions to Africa, Asia, North America, South America, and Central America.

151 Some critics take issue with this quote, arguing that Ali might have more naturally addressed Barbour in Dutch or Malay. I suggest there are several valid explanations. Undoubtedly Ali would have learned a bit of English while with Wallace (and with Spenser St. John and the Boyle brothers), and he was dredging out his limited English vocabulary to impress foreign visitors who obviously were interested in biology, particularly after having heard them speak English among themselves. Perhaps Ali *did* address Barbour in Dutch or Malay, and Barbour simply gave readers the English translation. And as for Ali referring to himself as Ali Wallace, well that makes sense, since Malays don't use family names but refer to themselves as the son or daughter of so-and-so. Why wouldn't Ali consider himself Wallace's son, or at least an adopted or honorary son? (And the common honorific throughout Indonesia given to an older man is *bapak*, or "father"). Perhaps the encounter was not even as theatrical as Barbour claims; he might have dramatized it for the sake of his narrative. But it might well have occurred as reported since we know Ali was a bit of a showman, as evidenced by Wallace's recollection of Ali's pleasure in telling exaggerated stories that enhanced his standing in the community. And it is unlikely any other local resident of Ternate, except Ali, would have had a serious discussion with a foreign collector about an obscure lizard. For me, there is no reason to doubt Barbour's sincerity in reporting his meeting with Ali.

152 Barbour, Thomas (1943).

153 It would have been nice if Wallace had also written a note to Ali.

154 I think it is important that Wallace expressed no surprise that Ali was living in Ternate, so his letter to Barbour is one more piece of evidence that Ali "retired" to Ternate. Of course, one might interpret Wallace's request for a photo of Ali as a very polite way of expressing doubt about the accuracy of Barbour's account and asking for proof.

155 For a discussion of this non-existent photograph, see Wyhe, John van, and Drawhorn, Gerrell (2015).

CHAPTER 14
The Search for Ali's Descendants

156 Wallace, Alfred Russel (1867).

157 For more on the search for the location of Wallace's house, see Hughes, Nicholas, and Taib, Rinto (2022) and Whincup, Paul (2020).

CHAPTER 15
Might the Ghosts of Wallace and Ali Help Me Unravel the Unanswered Questions about Ali's Life?

158 I am not the first to consider a form of History by Hantu. French author Victor Hugo claimed he had conversations with Mozart, Dante, Plato, Galileo, and Moses. Hugo said that Jesus visited him three times, during which he outlined a new religion with Hugo as its prophet. Hugo also channeled Shakespeare, who helped the Frenchman write a play featuring Heaven, Hell, Paradise, Louis XV, and a peasant maiden named Nihila. In one of the play's long soliloquies Paradise tells Hell: "How happy mankind is! No more evil! ... Mankind is an immense flower whose roots are bathed in light and who has as many petals as the mouth of God has kisses."

159 Another oft-cited case of spirits supporting writers: In the 1920s the Rev. John Lamond spent his holidays at the village house in Domremy, France, where Joan of Arc grew up. As recounted by Lamond's friend Graham Moffat, the reverend spent hours meditating beside the famous "fairy tree" where Joan saw her visions and heard the voices of her guides. Moffat wrote: "By this means, though he himself was devoid of psychic gifts, [Lamond] hoped to get into touch with the still living spirit of the 'Maid of Orleans.'" Back in London Lamond had numerous "interviews" with Joan through the trance mediumship of a psychic named Mrs. Mason, learning "much interesting information that was entirely unknown to her biographers." The result of this spirit-enhanced research: A biography, *Joan of Arc and England*, and a play about her life. "While it must be admitted that G.B.S. [George Bernard Shaw] has given us a drama more skilfully constructed and more intensely dramatic, Dr. Lamond's work depicts the real Joan," noted Moffat, a playwright himself who believed that his dead father and brother helped him write his plays. "When we consider that it was written by a Scottish parson whose knowledge of theatrical technique cannot have been very profound, [Lamond's] play is an astonishing fine piece of work. Here is no credulous village girl deceived by church bells into thinking that she hears voices; no victim of hallucinations; but a clairvoyant and clairaudient maid directly inspired from the spirit world and raised by Heaven-given power to be the saintly heroine of France ... divinely inspired, her campaign against the English invaders of her country was designed and carried out with the aid of her spirit guides."

160 Noted science writer David Quammen said: "Wallace was a man of crotchety independence and lurching enthusiasms. If he hadn't existed, it would have taken a very peculiar Victorian novelist to create him." Two "very peculiar" writers did just that: Joseph Conrad, who said *The Malay Archipelago* was his favorite bedside reading, modeled the naturalist Stein in *Lord Jim* on Wallace, about which historian Lloyd Fernando writes: "Through Stein's [Wallace's] orgasmic taxonomic discoveries, Conrad illustrates the gap between notions of human idealism and sordid reality." Another "very peculiar" writer was Arthur Conan Doyle, of Sherlock Holmes fame, who based the character of the naturalist Stapleton in *The Hound of the Baskervilles* on Wallace. So profound was Wallace's spirit on Conan Doyle that the author, who was a strong believer in spiritualism, noted that "an invisible and friendly presence" that provided him with literary advice was the ghost of Wallace.

161 In addition to basing fictional characters on Wallace himself, several noted writers have referred to *The Malay Archipelago* in their works. Joseph Conrad refers directly to what he calls "Alfred Wallace's famous book on the Malay Archipelago," in "The Secret Agent." In W. Somerset Maugham's short story, "Neil MacAdam," the title character reads *The Malay Archipelago* while travelling to Borneo, and Maugham's description of that vast island reflects Wallace's influence.
162 Wallace, Alfred Russel (1876a).
163 Wallace, Alfred Russel (1887).
164 Wallace, Alfred Russel (1876a).
165 Wallace, Alfred Russel (1876a).
166 Wallace, Alfred Russel (1876a)
167 Frederick Hudson's photographs were exposed as fraudulent by numerous investigators, see: McCabe, Joseph (1920), Price, Harry (1936), Edmunds, Simeon (1966), and Milbourne, Christopher (1975). He used a specially built trick camera, made by a craftsman who sold a conjuring apparatus which produced a single exposure with images of the sitter and a "ghost." Though frequently caught practicing deception, he was never arrested.
168 Roper Center for Public Opinion Research (2018).
169 Pew Research Center (2009).
170 These narratives are excerpted from my book *Dead, But Still Kicking*, which, in addition to attempted conversations with Wallace and Ali includes attempts to speak with a man-hating female vampire ghost in Borneo, a river nature spirit in Burma, and the mystical Mermaid Queen in Java, who is the consort of two important sultans.

Chapter 16
Good News! Alfred Russel Wallace Is My Spirit Guide, My Mentor, My Pal

171 Wallace, Alfred Russel (1875).
172 For Wallace's detailed description of the three spirit photos taken by Hudson, see *Miracles and Modern Spiritualism*.
173 Some of Wallace's interests and areas of expertise are well-known: natural history, biogeography, conservation, glaciology, ethnography, land nationalization, astrobiology (is there life on Mars?), and spiritualism. A wide-ranging campaigner, he argued in favor of the rights of women, a minimum wage, food and drug standards, and an income tax. With similar zeal, he argued against gambling, "red-tapism," air pollution, eugenics, child labor, vivisection, militarism, vaccination, and imperialism.
174 Both acids are considered humic substances; they are derived from vegetative matter and increase the microbial activity in the soil, making them excellent root stimulators. They are occasionally used as medicines, to stimulate the immune system.

Chapter 17
Spirit Conversations with the Elusive Ali

175 Just as Malaysia is rich in biodiversity, the country is also certainly in the top tier of the number of ghost species it is home to. For a wonderful encyclopedia of some 126 Malaysian ghosts, vampires, hantus, demons, were-tigers, goblins, and other creatures you don't want to meet on a dark and stormy night, see Danny Lim's *The Malaysian Book of the Undead*. My favorite: *orang minyak*.
176 For more on this phenomenon, called transmogrification, as it occurs in Southeast Asia, see McNeely, Jeffrey, and Sochaczewski, Paul Spencer. *Soul of the Tiger*. University of Hawaii Press. (1995).
177 Wallace, Alfred Russel (1869).
178 Wallace, Alfred Russel (1869).
179 The term likely is a loan word from the English "to haunt."
180 I speak reasonable Bahasa Indonesia but need a translator to understand subtleties and the sometimes-archaic dialects mediums use while in trance.

181 Trying to make a windfall profit from a gullible foreigner seems to be a regular practice among some of the mediums I visited. A medium in Pontianak, Indonesia, tried to fleece me when I asked him to channel a female vampire ghost, and a famous shaman in Yangon, Myanmar, quoted a ridiculous price to speak with a nature spirit (more details in *Dead, But Still Kicking*). Most "serious" mediums I met either do not charge (but will accept a gift) or accept a specific fee only if they are able to make contact.
182 Sultan of Ternate Mudaffar Syah of Ternate, Ofa Firman's father, reigned from 1975 to 2015.
183 Beccaloni, George (2019a).

References

Ashby, Jack (2020). "Telling the Truth About Who Really Collected the 'Hero Collections.'" National Sciences Collections Association. October 22, 2020. https://natsca.blog/2020/10/22/telling-the-truth-about-who-really-collected-the-hero-collections/

Barbour, Thomas (1912). "A Contribution to the Zoogeography of the East Indian Islands." *Memoirs of the Museum of Comparative Zoology at Harvard College*, Vol XIIV, No 1.

Barbour, Thomas (1921). "Aquatic Skincs and Arboreal Monitors." *Copeia*, No. 97 (Aug. 31), pp. 42–44.

Barbour, Thomas (1943). *Naturalist at Large*. Boston: Little, Brown.

Bates, Henry Walter (1863). *The Naturalist on the River Amazons*. London: John Murray.

Beccaloni, George (2017a). *Introduction to The Malay Archipelago*. London: Folio Society.

Beccaloni, George (2017b). "Plants and Animals Named After Wallace." The Alfred Russel Wallace Website. https://wallacefund.myspecies.info/plants-and-animals-named-after-wallace

Beccaloni, George (2018). "Alfred Russel Wallace – A Very Important Ornithologist." The Alfred Russel Wallace Website. https://wallacefund.myspecies.info/content/alfred-russel-wallace-very-important-ornithologist

Beccaloni, George (2019a). "Wallace's Collections." The Alfred Russel Wallace Website. https://wallacefund.myspecies.info/wallaces-specimens

Beccaloni, George (2019b). "Specimen of Wallace's Standardwing Bird of Paradise from Alfred Russel Wallace's Own Collection Acquired by Singapore Natural History Museum." The Alfred Russel Wallace Website. https://wallacefund.myspecies.info/content/specimen-wallaces-standardwing-bird-paradise-alfred-russel-wallaces-own-collection-acquired

Beddall, Barbara G. (1988). *Darwin and Divergence: The Wallace Connection. Journal of the History of Biology*. Vol. 21, No. 1. Springer.

Berry, Andrew (2003). *Infinite Tropics: An Alfred Russel Wallace Anthology*. London: Verso.

Boyle, Frederick (1865). *Adventures Amongst the Dyaks of Borneo*. London: Hurst & Blackett.

Boyle, Frederick (1876). *The Savage Life: A Second Series of "Camp Notes."* London: Chapman and Hall.

Brackman, Arnold (1980). *A Delicate Arrangement: The Strange Case of Charles Darwin and Alfred Russel Wallace*. New York: Times Books.

Brooks, John L. (1984). *Just Before the Origin: Alfred Russel Wallace's Theory of Evolution*. New York: Columbia University Press.

Campbell, Joseph (1949). *The Hero with a Thousand Faces.* New York: Pantheon.

Cranbrook, Earl of; Hills, Daphne M.; McCarthy, Colin; and Prys-Jones, Robert (2005). "A. R. Wallace, Collector: Tracing His Vertebrate Specimens." In *Wallace in Sarawak – 150 Years Later: Proceedings of an International Conference on Biogeography and Biodiversity.* A.A. Ten and Indraneil Das, editors. Samarahan, Malaysia: IBEC, UNIMAS.

Cranbrook, Earl of (2008). "Alfred Wallace, Field Collector." In *Natural Selection and Beyond: The Intellectual Legacy of Alfred Russel Wallace.* Smith, C.H., and Beccaloni, George, editors. Oxford: Oxford University Press.

Cranbrook, Earl of, and Marshall, Adrian (2014). "Alfred Russell Wallace's Assistants, and Other Helpers, in the Malay Archipelago 1854-62." *Sarawak Museum Journal*, 73 (ns 94), pp. 73–122.

Desjeux, Isabelle. "Buang, the Lost Malay Scientist." https://isabellecreates.wordpress.com/projects/buang-the-lost-malay-scientist/

Drawhorn, Gerrell M. (2017). "Ali the Cook and 'Peter' the Collector: Two Wallace Helpers in His Last Sarawak Days." *Sarawak Museum Journal* LXXVIII:99 (ns), pp. 139–156.

Drawhorn, Jerry (2016). "The Alienation of Ali: Was Wallace's Assistant from Sarawak or Ternate?" *Sarawak Museum Journal,* LXXVI:97 (ns), pp. 165–200.

Drawhorn, Jerry (2018). Personal correspondence.

Drawhorn, Jerry (2023). Personal correspondence.

Edmunds, Siméon (1966). *Spiritualism: A Critical Survey.* Charlottesville: Aquarian Press, University of Virginia.

Fichman, Martin (2004). *An Elusive Victorian: The Evolution of Alfred Russel Wallace.* Chicago: University of Chicago Press.

Grouw, Hein van (2020). "De Vogelcollectie van Alfred Russel Wallace in het Britse Natural History Museum." In: Holleman, Roos and Reeuwijk, Alexander, *Wallacea: Een Ode aan Alfred Russel Wallace.* Gorredijk: Noordboek.

Hughes, Nicholas, and Taib, Rinto (2022). "The Quest for the Legendary House of Alfred Russel Wallace in Ternate." In press.

Lim, Danny (2008). *The Malaysian Book of the Undead.* Kuala Lumpur: Matahari Books.

Marchant, James, editor (1916). *Alfred Russel Wallace Letters and Reminiscences.* London: Cassell.

McCabe, Joseph (1920). *Spiritualism: A Popular History from 1847.* New York: Dodd, Mead and Company.

McLaughlin, Tom, and Suriani binti Sahari (2017). "The Mysterious Ali of the 'Malay Archipelago.'" https://imaginedmalaysia.wordpress.com/2017/03/14/the-mysterious-ali-of-the-malay-archipelago/

Milbourne, Christopher (1975). *Mediums, Mystics & the Occult.* Charlottesville: Ty Crowell Co.

Moffat, Graham (1950). *Towards Eternal Day: The Psychic Memoirs of a Playwright.* London: Rider & Co. Excerpted in: Price, Leslie (2016). "Joan of Arc and Dr. Lamond." *Psypioneer.* Vol. 12, No. 3, May–June.

Pew Research Center (2009). "Many Americans Mix Multiple Faiths." https://www.pewresearch.org/religion/2009/12/09/many-americans-mix-multiple-faiths/#ghosts-fortunetellers-and-communicating-with-the-dead

Price, Harry (1936). *Confessions of a Ghost-Hunter.* London: Putnam & Co.

Raby, Peter (2002). *Alfred Russel Wallace: A Life.* Princeton: Princeton University Press.

Reece, Bob (1997). "The Loves of Hugh Low." *Borneo Research Bulletin.*

Ritchey, David (2003). *The H.I.S.S. of the A.S.P.: Understanding the Anomalously Sensitive Person.* Terra Alta: Headline Books.

Roper Center for Public Opinion Research (2018). "Paradise Polled: Americans and the Afterlife." Roper Center, Cornell University.

Rookmaaker, K., and Wyhe, John van (2012). "In Alfred Russel Wallace's Shadow: His Forgotten Assistant, Charles Allen (1839–1892)." *Journal of the Malaysian Branch of the Royal Asiatic Society.*

Shermer, Michael (2002). *In Darwin's Shadow: The Life and Science of Alfred Russel Wallace.* Oxford: Oxford University Press.

Sirlin, Avi (2014). *The Evolutionist.* Twickenham: Aurora Metro Books.

Smith, Charles H., and Beccaloni, George, editors (2008). *Natural Selection and Beyond: The Intellectual Legacy of Alfred Russel Wallace.* Oxford: Oxford University Press.

Smith, Charles H.; Costa, James T.; and Collard, David, editors. (2019). *An Alfred Russel Wallace Companion.* Chicago: The University of Chicago Press.

Smith, Charles H. (2022). "Alfred Russel Wallace Notes 20: Did Darwin and Wallace 'Coauthor' the 1858 Communication on Natural Selection?" The Alfred Russel Wallace Page. https://works.bepress.com/charles_smith/91/

Sochaczewski, Paul Spencer (2015). "The God Who Flew Off with a Mountain." *Curious Encounters of the Human Kind — Himalaya.* Geneva: Explorer's Eye Press.

Sochaczewski, Paul Spencer (2017). *An Inordinate Fondness for Beetles.* Second edition. Geneva: Explorer's Eye Press.

Sochaczewski, Paul Spencer (2019). *Dead, But Still Kicking.* Geneva: Explorer's Eye Press.

Sochaczewski, Paul Spencer, and Hallmark, David Spencer (2005). "Alfred Russel Wallace: On the Origin of a Theory." *Geographical* (The magazine of the Royal Geographical Society). December.

Stevenson, Matthew (2023). "The Great Santos." In *Counterpunch*, January 20.

St. John, Spenser (1863). *Life in the Forests of the Far East.* London: Smith, Elder and Co.

St. John, Spenser (1879). *The Life of Sir James Brooke, Rajah of Sarawak, From His Personal Papers and Correspondence.* Edinburgh and London: W. Blackwood and Sons.

Storr, Glen M. 1971. "The genus Lerista (Lacertilia: Scincidae) in Western Australia." *Journal of the Royal Society of Western Australia*, Vol. 54, pp. 59–75.

Vogler, Christopher. (1992) *The Writer's Journey: Mythic Structure for Storytellers & Screenwriters.* Studio City, California: Michael Wiese Productions.

Wallace, Alfred Russel (1852). "Letter Concerning the Fire on the 'Helen.'" Letter to the Editor in "Proceedings of Natural-History Collectors in Foreign Countries" section of the *Zoologist* issue of November 1852. Published in The Alfred Russel Wallace Page. https://people.wku.edu/charles.smith/wallace/S007.htm

Wallace, Alfred Russel (1854a). Letter to mother from Singapore dated April 30, 1854.

Wallace, Alfred Russel (1854b). Letter to mother from Singapore dated September 30, 1854.

Wallace, Alfred Russel (1855). Letter to sister from Sarawak dated June 25, 1855.

Wallace, Alfred Russel (1856a) Letter to sister and brother-in-law from Singapore dated February 20, 1856.

Wallace, Alfred Russel (1856b). "On the Habits of the Orang-Utan of Borneo." *Annals & Magazine of Natural History*, July.

Wallace, Alfred Russel (1859a). Letter to John Gould dated 30 September 1859, published in *Proceedings of the Zoological Society of London for 1860.* http://people.wku.edu/charles.smith/wallace/S055.htm.

Wallace, Alfred Russel (1859b). Letter to Samuel Stevens, from Awaiya, Ceram. Nov. 26.

Wallace, Alfred Russel (1859c). "On the Zoological Geography of the Malay Archipelago." Paper presented to the Linnean Society on November 3, 1859, and published in their *Zoological Proceedings in 1860.*

Wallace, Alfred Russel (1863). "On the Physical Geography of the Malay Archipelago" [A paper read at the RGS meeting of 8 June 1863]. *Journal of the Royal Geographical Society of London.*

Wallace, Alfred Russel (1866). *The Scientific Aspect of the Supernatural: Indicating the Desirableness of an Experimental Enquiry by Men of Science into the Alleged Powers of Clairvoyants and Mediums.* London: F. Farrah.

Wallace, Alfred Russel (1869). *The Malay Archipelago: The Land of the Orang-Utan, and the Bird of Paradise, a Narrative of Travel, with Studies of Man and Nature.* 2 vols. London: Macmillan and Co.

Wallace, Alfred Russel (1874). "A Defence of Modern Spiritualism." *Fortnightly Review*, published in two parts: May and June.

Wallace, Alfred Russel (1876a). *On Miracles and Modern Spiritualism.* London: James Burns.

Wallace, Alfred Russel (1876b). *The Geographical Distribution of Animals.* London: Macmillan and Co.

Wallace, Alfred Russel (1893). "Notes on the Growth of Opinion as to Obscure Psychical Phenomena During the Last Fifty Years." *Religio-Philosophical Journal* (ns) 4, No. 15, pp. 440–41.

Wallace, Alfred Russel (1896). *Miracles and Modern Spiritualism.* London: George Redway.

Wallace, Alfred Russel (1887). "If a Man Die Shall He Live Again?" Lecture at Metropolitan Temple. San Francisco. June 5, 1887.

Wallace, Alfred Russel (1905). *My Life: A Record of Events and Opinions.* 2 vols. London: Chapman & Hall.

Whincup, Paul (2020). "The Quest for Alfred Russel Wallace's House on Ternate, Maluku Islands, Indonesia." *Journal of the Royal Society of Western Australia.* No. 103, pp. 50–54.

Wilson, John G. (2000). *The Forgotten Naturalist: In Search of Alfred Russel Wallace.* Kew, Victoria: Australian Scholarly Publishing Pty. Ltd.

Woodard, Colin (2023) "How the Myth of the American Frontier Got Its Start." *Smithsonian*, January–February.

Wyhe, John van, and Rookmaaker, Kees, editors (2013). *Alfred Russel Wallace: Letters from the Malay Archipelago.* Oxford: Oxford University Press.

Wyhe, John van (2013). *Dispelling the Darkness: Voyage in the Malay Archipelago and the Discovery of Evolution by Wallace and Darwin.* Singapore: World Scientific Publishing Company.

Wyhe, John van, and Drawhorn, Gerrell M. (2015). "'I am Ali Wallace': The Malay Assistant of Alfred Russel Wallace." *Journal of the Malaysian Branch of the Royal Asiatic Society.* 88:1:3–31.

Wyhe, John van (2018). "Wallace's Help: The Many People Who Aided A. R. Wallace in the Malay Archipelago." *Journal of the Malaysian Branch of the Royal Asiatic Society.* June 2018, Vol. 91, Part 1, No. 314.

Further Reading and Key Resources

Background, Commentary and Important Reference Sites

Alfred Russel Wallace Page. Charles H. Smith, editor. The modest name of this extensive site does not do justice to the expansive references, texts, and commentaries about Wallace. Among other resources, the site includes a complete bibliography of Wallace's writing and an extensive bibliography of writing about Wallace. A section that I find particularly interesting contains Wallace obituaries, which reveal how respected, even lionized, he was at his death. Invaluable. https://people.wku.edu/charles.smith/index1.htm#gsc.tab=0.

The Alfred Russel Wallace Website. Created and curated by George Beccaloni; the official website of the Wallace Memorial Fund. A treasure chest of essential information, anecdotes, curious facts, and opinions. https://wallacefund.myspecies.info/.

Beccaloni, George, editor (2023). *Catalogue of the Correspondence of Alfred Russel Wallace.* First edition. DOI: 10.5281/zenodo.7769858. Also available at: https://zenodo.org/record/7769858#.ZCAZBXZBxD8. This publication contains records of 5,688 letters, of which 2,748 were written by Wallace and 2,159 were sent to him. The remaining 781 are third-party letters that either pertain to him or are written by relatives, which contain information useful to scholars interested in Wallace's life. Also available at the Alfred Russel Wallace Correspondence Project: https://wallaceletters.myspecies.info/content/wallace-correspondence-project-insights-remarkable-life.

Drawhorn, Jerry (2016). "The Alienation of Ali: Was Wallace's Assistant from Sarawak or Ternate?" *Sarawak Museum Journal.* A thorough review of Wallace's linguistic skills and interests, and a useful review of the use of the Malay language in the Malay Archipelago, particularly as it relates to the question of whether Ali originated in Sarawak or Ternate.

Natural History Museum (London) Wallace Collection. The museum's collections include some 5,000 Wallace-related specimens, photographs, and documents. https://www.nhm.ac.uk/discover.html.

Wallace Online. The first complete edition of the writings of Alfred Russel Wallace, including the first compilation of his specimens. http://wallace-online.org/.

Birds-of-paradise

The Cornell Lab of Ornithology is a valuable resource center for data, photos, and videos of all the birds of paradise, including Wallace's standardwing. https://ebird.org/species/walsta2.

Dickinson, Edward C. and Christidis, Les, editors (2014). *The Howard and Moore Complete Checklist of the Birds of the World, Volume 2.* Totnes: NHBS-Aves Press.

Lawrence, Natalie. (2018). "Fallen Angels: Birds of Paradise in Early Modern Europe." In *The Public Domain Review*. This article offers a useful guide to the historic European infatuation with birds of paradise. https://publicdomainreview.org/essay/fallen-angels-birds-of-paradise-in-early-modern-europe.

Linn-Gutowski, Melanie, and Docent, Clayton (2021). *Crimes of Fashion: Gilded Age Millinery and the Plight of Birds*. Pittsburgh: The Frick Pittsburgh. Provides a history of birds of paradise in Victorian fashion. https://www.thefrickpittsburgh.org/Story-Crimes-of-Fashion-Gilded-Age-Millinery-and-the-Plight-of-Birds.

Johnson, Kirk Wallace (2018). *The Feather Thief.* New York: Viking. https://www.amazon.com/Feather-Thief-Obsession-Natural-History/dp/1101981636/.

Smith, Malcolm (2020). *Hats: A Very UNnatural History*. East Lansing: Michigan State University Press. https://www.amazon.com/Hats-Very-UNnatural-History-Animal/dp/1611863473.

Alfred Russel Wallace's Books (Selection)

Several of Wallace's more interesting/important publications (there are many more — he wrote 21 books, more than 600 academic papers and articles, and hundreds of letters). His scope covers subjects as diverse as evolution, biography, island biology, ethnography, conservation, spiritualism, and astrobiology. He also wrote extensively and campaigned about social issues, including the rights of women, militarism, colonial policy, child labor, air pollution, food and drug standards, and land nationalization. Not bad for a lad who left school at the age of 14.

Wallace, Alfred Russel (1853). *A Narrative of Travels on the Amazon and Rio Negro: With an Account of the Native Tribes, and Observations on the Climate, Geology, and Natural History of the Amazon Valley*. London: Reeve & Co.

Wallace, Alfred Russel (1869). *The Malay Archipelago: The Land of the Orang-Utan, and the Bird of Paradise, a Narrative of Travel, with Studies of Man and Nature.* 2 vols. London: Macmillan & Co.

Wallace, Alfred Russel (1876a). *On Miracles and Modern Spiritualism*. London: James Burns.

Wallace, Alfred Russel (1876b). *The Geographical Distribution of Animals; with a Study of the Relations of Living and Extinct Faunas as Elucidating the Past Changes of the Earth's Surface.* London: Macmillan & Co.

Wallace, Alfred Russel (1880). *Island Life, or the Phenomena and Causes of Insular Faunas and Floras, including a Revision and Attempted Solution of the Problem of Geological Climates.* London: Macmillan & Co.

Wallace, Alfred Russel (1886). *Bad Times: An Essay on the Present Depression of Trade, Tracing It to Its Sources in Enormous Foreign Loans, Excessive War Expenditure, the Increase of Speculation and of Millionaires, and the Depopulation of the Rural Districts; With Suggested Remedies*. London: Macmillan & Co.

Wallace, Alfred Russel (1898). *The Wonderful Century; Its Successes and Its Failures*. London: Swan Sonnenschein & Co.

Wallace, Alfred Russel (1889). *Darwinism: An Exposition of the Theory of Natural Selection with Some of Its Applications*. London: Macmillan & Co.

Wallace, Alfred Russel (1903). *Man's Place in the Universe: A Study of the Results of Scientific Research in Relation to the Unity or Plurality of Worlds*. London: Chapman & Hall.

Wallace, Alfred Russel (1905). *My Life; A Record of Events and Opinions*. 2 volumes. London: Chapman & Hall.

Wallace, Alfred Russel (1907). *Is Mars Habitable? A Critical Examination of Professor Percival Lowell's Book "Mars and Its Canals," with an Alternative Explanation*. London: Macmillan & Co.

Other Interesting References

A very selective list of miscellaneous publications about conservation, Southeast Asia, taxidermy, and culture that indirectly relate to Wallace and Ali.

Bangs, Richard, and Kallen, Christian. (1988). *Islands of Fire, Islands of Spice*. San Francisco: Sierra Club Books.

Barley, Nigel (2002). *White Rajah*. New York: Little Brown.

Birds: Instructions for Collectors No 2A. (1970). London: Trustees of the British Museum (Natural History).

Blair, Lawrence, with Blair, Lorne (1988). *Ring of Fire*. New York: Bantam Books.

Hannigan, Tim. (2015). *A Brief History of Indonesia*. Singapore: Tuttle.

Katz, Daniel R., and Chapin Miles, editors (1995). *Tales from the Jungle: A Rainforest Reader*. New York: Crown.

King, Victor T., editor (1992). *The Best of Borneo Travel*. Oxford: Oxford in Asia.

McNeely, Jeffrey, and Sochaczewski, Paul Spencer (1991). *Soul of the Tiger*. Oxford: Oxford University Press.

Miller, George, editor (1996). *To the Spice Islands and Beyond.* Oxford: Oxford University Press.

Milton, Giles (1999). *Nathaniel's Nutmeg.* London: Hodder and Stoughton.

O'Hanlon, Redmond (1984). *Into the Heart of Borneo.* Edinburgh: The Salamander Press.

Sochaczewski, Paul Spencer (2015). *Curious Encounters of the Human Kind – Indonesia.* Geneva: Explorer's Eye Press.

Sochaczewski, Paul Spencer (2015). *Curious Encounters of the Human Kind – Southeast Asia.* Geneva: Explorer's Eye Press.

Sochaczewski, Paul Spencer (2015). *Curious Encounters of the Human Kind – Borneo.* Geneva: Explorer's Eye Press.

Sochaczewski, Paul Spencer (2016). *Redheads.* Geneva: Explorer's Eye Press.

Sochaczewski, Paul Spencer (2018). *Exceptional Encounters: Enhanced Reality Tales from Southeast Asia.* Geneva: Explorer's Eye Press.

Sochaczewski, Paul Spencer (2020). *EarthLove.* Geneva: Explorer's Eye Press.

Sochaczewski, Paul Spencer (2022). *A Conservation Notebook.* Geneva: Explorer's Eye Press.

Spiritualism and Science

Barušs, Imants, and Mossbridge, Julia (2016). *Transcendent Mind: Rethinking the Science of Consciousness.* Washington, DC.: American Psychological Association.

Conan Doyle, Arthur (1926). *The History of Spiritualism.* London: Cassell & Company.

Delorme, Arnaud; Beischel, Julie; Michel, Leena; Boccuzzi, Mark; Radin, Dean; Mills, Paul (2013). "Electrocortical Activity Associated with Subjective Communication with the Deceased." *Frontiers in Psychology.* November 20, Vol. 4, article 834.

Greeley, Andrew M., and Hout, Michael (1999). "Americans' Increasing Belief in Life After Death: Religious Competition and Acculturation." *American Sociological Review* 64(6), 813–835.

Sochaczewski, Paul Spencer (2019). *Dead, But Still Kicking: Encounters with Mediums, Shamans, and Spirits.* Geneva: Explorer's Eye Press.

Taylor, Steve (2018). *Spiritual Science: Why Science Needs Spirituality to Make Sense of the World.* London: Watkins Publishing.

Wallace, Alfred Russel (1866). *The Scientific Aspect of the Supernatural: Indicating the Desirableness of an Experimental Enquiry by Men of Science into the Alleged Powers of Clairvoyants and Mediums*. London: F. Farrah.

My spirit conversation with Wallace was enabled via soul artist and clairvoyant June-Elleni Laine, who also asked her students for their psychic impressions of Ali. She runs spirit workshops and consultations, contact her at: https://www.june-elleni.com/

Geneva-based Brigitte Favre can be reached at: https://www.psycho-ge.ch/?lang=en (psychological counseling) and https://www.2-worlds.ch/?lang=en#contact (mediumship).

Moïra Salvadore, who offered her thoughts about Ali based on his photo, teaches intuition in Geneva; she can be reached at: www.intuition-in-the-city.com

Praise for
Paul's Other Books

A Conservation Notebook

Ego, Greed, and Oh-So-Cute Orangutans — Tales from a Half-Century On the Environmental Front Lines

Explorer's Eye Press, Geneva, 2022
ISBN: 978-2-940573-39-4

This highly personal volume from the former head of Creative Services for WWF International wanders the world in search of conservation successes and failures, heroes and villains. The book contains no finger-wagging lectures, not too many depressing statistics, and no easy solutions. It is a collection of curious encounters and outlying ideas reflecting five decades of work in the nature conservation wonderland, linked by the theme that nature is too important to ignore.

"An endearing inspiration and a wonderful tribute to the movements and characters behind modern conservation."

Tobgay Sonam Namgyal, former head of the Bhutan Trust Fund for Environmental Conservation

"A must-read for all those involved in the race to save the planet."

Mark Halle, former director of IUCN, founding director of Better Nature

"Don't read this book seeking glib solutions. But do read it, please, to get a human, and frequently moving story about how conservation works in the real world."

James Clad, former bureau chief South and Southeast Asia, Far Eastern Economic Review, former U.S. deputy assistant secretary of defense for Asia

SEARCHING FOR GANESHA

**COLLECTING IMAGES OF THE SWEET-LOVING,
ELEPHANT-HEADED HINDU DEITY EVERYBODY ADMIRES**

Explorer's Eye Press, Geneva, 2021
ISBN: 978-2-940573-37-0

Ganesha, the Hindu elephant-headed god, is among the most-treasured of all deities. In this innovative book Sochaczewski explores why he collects Ganesha images (some 80 objects from his collection are shown in museum-quality photos), examines the psychology of collecting, and recounts personal adventures in his 40-year quest for just-one-more Ganesha statue. He describes the book as "a personal travel adventure with zero religious intent."

"A treasure. Part intellectual homage, part personal journey, part sheer whimsy. A noble tip of the hat to one of the world's favorite gods."

– Ro King, chair, Global Heritage Fund and chair emeritus, Indonesian Heritage Society

"Sochaczewski's deep knowledge is matched only by his humanistic spirit, humor, and modesty that runs through the book. Ganesha aside (because I will readily admit that the author's mastery of the subject is far greater than mine), his acute questioning of the desire to collect should become mandatory reading for budding museum professionals."

– Danien Kunik, curator for Asian collections, MEG-Musée d'ethnographie de Genève

EarthLove

Chronicles of the Rainforest War

Explorer's Eye Press, Geneva, 2020
ISBN: 978-2-940573-34-9

EarthLove, a satiric Borneo eco-adventure, chronicles the history of the global conservation movement and exposes battles pitting ego and greed versus noble intentions. Who has the power to stop the rape of the tropical rainforests? Is there hope for the people of the rainforest? For the orang-utans? For the forces of good to outlast the armies of evil?

"Scents of Carl Hiaasen, Edward Abbey, and Tom Wolfe combined into a unique voice of darkly comic fictional truth."

—*Simon Lyster, chairman, Conservation International, UK*

"An absolute delight, *EarthLove* reveals the dark, and deliciously satirical, underbelly of modern conservation."

– *Nigel Barley, author of* The Innocent Anthropologist,
former curator for Southeast Asia, British Museum

Dead, But Still Kicking

Encounters with Mediums, Shamans, and Spirits

Explorer's Eye Press, Geneva, 2019
ISBN: 978-2-940573-32-5

In this innovative work of personal journalism, Sochaczewski — a self-described Skeptical Spiritualist — creates the Three Tenets of Spiritualism while travelling to Indonesia, Myanmar, the United Kingdom, and Switzerland to speak with spirits of dead folks. He gets a personal mandate from Moses, speaks with Alfred Russel Wallace about his relationship with Charles Darwin, encounters a vengeful female vampire ghost, and converses with nature spirits.

"Enlightening. A noble companion volume to my own books on spiritualism."

— *Spirit of Arthur Conan Doyle*

"A brave attempt to understand the widening gyres."

— *Spirit of W.B. Yeats*

Exceptional Encounters

Enhanced Reality Tales from Southeast Asia

Explorer's Eye Press, Geneva, 2018

ISBN: 978-2-940573-29-5

Exceptional Encounters takes the seeds of true events and applies the classic fiction writer's aerobic exercise by asking: What if? These enhanced-reality fabulations draw the reader into tales of just over-the-rainbow Asian kindness, greed, passion, and dreams.

"A touch of George Orwell for our challenging times."

—Robin Hanbury-Tenison, founder of Survival International

"At turns outrageous, thoughtful, and darkly satirical. Pushes the frontier of personal travel literature into a new dimension."

—Simon Lyster, chairman World Land Trust

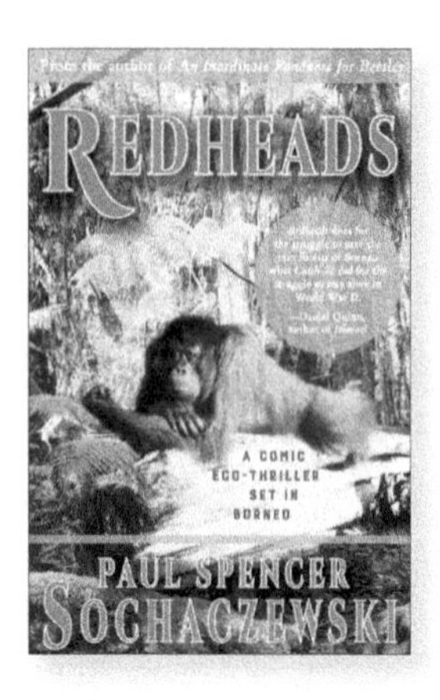

Redheads

A Comic Eco-Thriller Set in Borneo

Explorer's Eye Press, Geneva, 2016
ISBN: 978-2-940573-18-9

In the middle of a Borneo rainforest, a band of near-naked Penan tribesmen, encouraged by a similarly clothes-challenged renegade Swiss shepherd, blockade a logging truck. Nearby, a researcher studying orangutans is threatened with being thrown out of her study site unless she can reach a delicate compromise with the powerful minister of the environment. Meanwhile, a few identity-confused orangutans seek their own methods of survival.

Will the threatened homeland of people and orangutans survive?

"*Redheads* does for the struggle to save the rainforests of Borneo what *Catch-22* did for the struggle to stay alive in World War II."

—*Daniel Quinn, author of* Ishmael

"A visceral jungle morality play. Free-thinking, intelligent, and irreverent, reminds me of a Kurt Vonnegut thriller."

— *Benedict Allen, author of* Into the Crocodile's Nest: Journey Inside New Guinea

An Inordinate Fondness for Beetles

Campfire Conversations with Alfred Russel Wallace

Explorer's Eye Press, Geneva, 2017
ISBN: 978-2-940573-25-7

An Inordinate Fondness for Beetles follows the Victorian-era explorations of Alfred Russel Wallace through Southeast Asia.

Sochaczewski examines themes about which Wallace cared deeply and interprets them through his own filter with layers of humor, history, social commentary, and sometimes outrageous personal tales.

"The rhythm and magic of a verbal fugue. A new category of nonfiction—part personal travelogue, part incisive biography, part unexpected traveller's tales."

—*Dato Sri Gathorne, author of* Mammals of Borneo

"As if I had boarded a time machine. A revelation of Wallace's insights interwoven with Sochaczewski's unique view of the world and our place in it."

—*Thomas E. Lovejoy, professor at George Mason University, president of the Amazon Biodiversity Center*

Share Your Journey

Mastering Personal Writing

Explorer's Eye Press, Geneva, 2016

ISBN: 978-2-940573-15-8

Share Your Journey is an easy-to-use handbook for people who want to write their personal stories. The book's Ten Writing Tips gives writers the techniques professional authors use to write memoirs and travel stories that connect with readers and editors.

"This is a lifetime's wisdom, offered by a pro. Put *Share Your Journey* next to *The Elements of Style* by Strunk and White—they'll be the only two writing books you'll need."

—*Thomas Bass, author of* The Spy Who Loved Us

"*Share Your Journey* is to good writing as *Joy of Cooking* is to good food. It's smart, fun, and every page contains nuggets of essential advice."

— *Gary Goshgarian, professor of creative writing, Northeastern University*

Curious Encounters of the Human Kind

True Asian Tales of Folly, Greed, Ambition, and Dreams

Explorer's Eye Press, Geneva, 2016

A five-volume series—Myanmar (Burma), Southeast Asia, Indonesia, Himalaya, and Borneo — containing true stories based on Sochaczewski's 50 years of living and exploring in curious corners of Asia. This is Asia as you've probably never imagined, full of memorable people, startling happenings, and unexpected moments of humanity and introspection, giddiness and solemnity, avarice and ambition.

"The spirit of Kipling in contemporary Asian journalism. This collection is essential reading for anyone who wishes to pass beyond even the unbeaten track, right to the heart of Asia."

—*John Burdett, author of* Bangkok Asset

"The humanity of Somerset Maugham, the adventure of Joseph Conrad, the perception of Paul Theroux, and a self-effacing voice unique his own."

—*Gary Braver, bestselling author of* Tunnel Vision

Distant Greens

Golf, Life, and Surprising Serendipity On and Off the Fairways

Explorer's Eye Press, Geneva, 2016
ISBN: 978-2-940573-21-9

Distant Greens travels to the highest golf course in the world, where breathless Tibetan precepts come face to face with the Indian military. To a golf course in the Amazon rainforest, near the source of rubber, which revolutionized the game. To the Middle Kingdom, to examine claims that it was the Chinese who invented golf.

More than an insightful personal travelogue, *Distant Greens* also delves into the soul of the sport and shows how golf can be a force for nature conservation.

"An intimate golfing tour that travels to all corners of the planet and brings us into the heart, mind, and soul of the game that we all love."

—*Rick Lipsey,* Sports Illustrated

Soul of the Tiger

Searching for Nature's Answers in Southeast Asia

Jeffrey A. McNeely and Paul Spencer Sochaczewski

University of Hawai'i Press, Honolulu, 1995
ISBN: 0-82481-669-2

One recent reviewer noted: "Age has not diminished the value of this book; it remains a classic in the genres of both conservation and travel literature."

Soul of the Tiger identifies the four "eco-cultural revolutions" that have dramatically changed the face of Southeast Asia and suggests a fifth revolution that could lead to a new sustainable relationship between people and nature.

"One revealing, insightful, and stimulating account after another, focusing on the relationship between our own and other species. Importantly, it reveals why traditional human-wildlife relations should be encouraged in a world that seeks to balance economic growth and environmental preservation."

—*John Noble Wilford in* The New York Times

Eco-Bluff Your Way to Greenism

The Guide to Instant Environmental Credibility

Paul Spencer Sochaczewski
(writing as Paul Spencer Wachtel)
and Jeffrey A. McNeely

Bonus Books, Chicago, 1991
ISBN: 0-929387-22-8

The guide to attain quick and painless eco-credibility, with essential advice on things such as how to deal with people who prefer elephants to human beings, how to establish your street-cred by explaining the public relations coup of Chief Seattle, and how to stir up a party by roaring like an eco-guerilla.

"What a book! Covers insights into potentially disastrous global issues in a bright and enjoyable way. Takes no prisoners and opens our eyes to a new and more effective vision of the pathway to environmental sanity."

—Noel Vietmeyer, US National Academy of Sciences

The Sultan and the Mermaid Queen

Surprising Asian People, Places, and Things that Go Bump in the Night

Editions Didier Millet, Singapore, 2008
ISBN: 978-981-4217-74-3

These 70 true, unnerving, off-the-radar Asian tales confirm Sochaczewski's unique voice as one of the leading travel writers of his generation. Why do Javanese sultans owe their power to the Mermaid Queen? Why are Indian villagers angry at the Monkey God Hanuman for not returning their sacred mountain? Why is the Indonesian island of Flores ground zero for "small people" fables? And why was the 90-year-old "last elephant hunter" of Vietnam offered a lucrative product endorsement?

"Sochaczewski is a world-class searcher, reporter, and observer... an insightful guide to an often obscure and rapidly changing world."

—*Christopher G. Moore, author of the Vincent Calvino novels*

"That rarest of writers—he has discovered an eternal assemblage of arcane explorers, putative emperors, frivolous mystics, sacrosanct elephants, and yes, miracle workers."

— *Harry Rolnick, author of* Spice Chronicles: Exotic Tales of a Hungry Traveler

About the Author

The tale of Alfred Russel Wallace and Ali has intrigued me for more than 50 years. I've trailed Wallace up the Amazon and Rio Negro rivers and followed Wallace and Ali throughout Southeast Asia. I've written about them in dozens of magazine articles and in two books: *An Inordinate Fondness for Beetles* and *Dead, But Still Kicking.*

The Wallace-Ali relationship has many facets. Wallace was Ali's employer, but he was also the young man's teacher, protector, nurse, and father figure. In a strange way he was also probably a role model for the young man, who under Wallace's wing travelled the breadth of Southeast Asia. Likewise, Ali supported Wallace and no doubt taught him some useful survival lessons.

I welcome readers' thoughts about your own Wallace-Ali-like connections. To whom do you owe a debt of gratitude for helping you understand the direction of your life journey? And who have *you* helped explore the world?

William J. Stone

Paul reading *The Malay Archipelago* in a river in Sarawak not far from where Wallace first met Ali. He can be contacted at his website: www.sochaczewski.com

Milton Keynes UK
Ingram Content Group UK Ltd.
UKHW011819191023
430941UK00004B/46